BIG DATA

CONCEPTS, WAREHOUSING, AND ANALYTICS

RIVER PUBLISHERS SERIES IN INFORMATION SCIENCE AND TECHNOLOGY

Series Editors:

K. C. Chen
National Taiwan University, Taipei, Taiwan
and
University of South Florida, USA

Sandeep Shukla
Virginia Tech, USA
and
Indian Institute of Technology Kanpur, India

Indexing: All books published in this series are submitted to the Web of Science Book Citation Index (BkCI), to SCOPUS, to CrossRef and to Google Scholar for evaluation and indexing.

The "River Publishers Series in Information Science and Technology" covers research which ushers the 21st Century into an Internet and multimedia era. Multimedia means the theory and application of filtering, coding, estimating, analyzing, detecting and recognizing, synthesizing, classifying, recording, and reproducing signals by digital and/or analog devices or techniques, while the scope of "signal" includes audio, video, speech, image, musical, multimedia, data/content, geophysical, sonar/radar, bio/medical, sensation, etc. Networking suggests transportation of such multimedia contents among nodes in communication and/or computer networks, to facilitate the ultimate Internet.

Theory, technologies, protocols and standards, applications/services, practice and implementation of wired/wireless networking are all within the scope of this series. Based on network and communication science, we further extend the scope for 21st Century life through the knowledge in robotics, machine learning, embedded systems, cognitive science, pattern recognition, quantum/biological/molecular computation and information processing, biology, ecology, social science and economics, user behaviors and interface, and applications to health and society advance.

Books published in the series include research monographs, edited volumes, handbooks and textbooks. The books provide professionals, researchers, educators, and advanced students in the field with an invaluable insight into the latest research and developments.

Topics covered in the series include, but are by no means restricted to the following:

- Communication/Computer Networking Technologies and Applications
- Queuing Theory
- Optimization
- Operation Research
- Stochastic Processes
- Information Theory
- Multimedia/Speech/Video Processing
- Computation and Information Processing
- Machine Intelligence
- Cognitive Science and Brain Science
- Embedded Systems
- Computer Architectures
- Reconfigurable Computing
- Cyber Security

For a list of other books in this series, visit www.riverpublishers.com

BIG DATA

CONCEPTS, WAREHOUSING, AND ANALYTICS

MARIBEL YASMINA SANTOS

CARLOS COSTA

River Publishers

Published, sold and distributed by:
River Publishers
Alsbjergvej 10
9260 Gistrup
Denmark

www.riverpublishers.com

ISBN: 978-87-7022-184-9 (Hardback)
       978-87-7022-183-2 (Ebook)

Pagination: Alice Paula Simões
Cover design: José M. Ferrão – *Look-Ahead*

River Publishers Series in Information Science and Technology

All our books have a strict quality control, however we advise you to periodically check our website (www.fca.pt) to download any corrections.

We are not responsible for any error in the hyperlinks of this book, which were checked at publication date.

All trade names refered in this book have registered patent.

To the ones who are the music of my life.

Maribel Yasmina Santos

To everyone who believed in me and helped me keep going.

Carlos Costa

CONTENTS

LIST OF FIGURES

© FCA

© FCA

© FCA

LIST OF TABLES

© FCA

THE AUTHORS

Maribel Yasmina Santos

Associate Professor at the Department of Information Systems, University of Minho (http://www.uminho.pt), Senior Researcher of the ALGORITMI Research Centre (http://algoritmi.uminho.pt), and leader of SEMAG, the Software-based Information Systems Engineering and Management Group, at ALGORITMI. Her research interests include Business Intelligence and Analytics, Big Data Analytics, (Big) Data Warehousing and Online Analytical Processing. Coauthor of the book *Business Intelligence – Da Informação ao Conhecimento (3.ª Ed. At.)*, also published by FCA.

Carlos Costa

Invited Lecturer in the field of Information Systems, at the University of Minho (http://www.uminho.pt). Previous experiences include senior Big Data engineer, researcher, and software developer. (Co)author of several scientific and technical publications in the areas of Big Data, Data Warehousing, and Data Science. He is constantly looking for new ways of contributing to the community of researchers and practitioners in these areas.

ACKNOWLEDGMENTS

It was such a challenging and amazing journey the one that drove us to this book. During this journey, we received the help and support of many institutions, colleagues, friends, and family. It is now the time to thank you all, recognizing your valuable support and help. In particular, our special acknowledgments to:

- The University of Minho, our *home* for so many years. Thank you for giving us the resources, the problems, the students, and the context to carry out this project. A special recognition to the Department of Information Systems, to all our colleagues there, for supporting our enthusiasm and our work in this research field, to the ALGORITMI Research Centre, and *Centro de Computação Gráfica*;

- The various organizations that gave us so many challenges, allowing us to work on them and advance the state of the art in this field. We are particularly grateful to BOSCH Car Multimedia Portugal in Braga, for the excellent partnership and for providing the context for maintaining a remarkable working team;

- Our partners in the SusCity Project, and the many scientific and technical challenges we faced in such a multidisciplinary project;

- The amazing group of LID4/Lynx Lab, researchers that contributed every day to the work materialized in this book, addressing different research lines, ideas, projects, and discussions. Many thanks to João Galvão, Carina Andrade, Bruno Martinho, Francisca Lima, Eduarda Costa,

Guilherme Moreira, José Correia, João Martins, Cristiano Miranda, Fábio Teixeira, and Nuno Silva. Without you, this project would not have been possible. Thank you for the great work, but also for the lunches, dinners, cakes, and happiness we shared throughout these years;

- Our friends, a fundamental pillar in our lives;

- Our families, who are always by our side, supporting, encouraging, and giving us the strength to carry on;

- The open source community who provides valuable access to state-of-the-art technologies. Otherwise, research and development of most institutions would not be possible. Furthermore, we also acknowledge the relevance of people/ entities who provide free access to useful resources, such as Freepick from www.flaticon.com that made available icons included in some figures of this book.

FOREWORD

These are exciting times for our database community, from theory to practice, with many emerging and changing technologies trying to keep pace with the exponential data growth and heterogeneity of data sources, especially as we go through the evolving challenges of building scale-out architectures and complex analytical queries. *Big Data: Concepts, Warehousing, and Analytics* presents this diverse and shifting landscape comprehensively and thoroughly, while remaining accessible and welcoming to students, instructors, practitioners, and researchers.

It is not necessary for the readers to have an extensive knowledge of Big Data Warehousing (BDWing) to benefit from this book. It focuses on well-defined methods for designing, modeling, and querying to develop data flowing through different logical/physical components as well as analytical applications. The chapters have been written with a particular eye toward BDWing, and they meet the right balance between fundamental concepts and best practices related to the state of the art in Big Data and the multidisciplinary field of Data Warehousing (DWing). Many chapters contain suggestions for technological infrastructures as well as data modeling guidelines, offering the readers a unique view of BDWing performance for fast retrieval, *ad hoc* query granularities, and advanced analytical algorithms.

The authors of this book, Maribel Yasmina Santos and Carlos Costa, have contributed to the BDWing field for many years. Maribel has worked as a researcher, educator, and project leader. She has contributed for many decades in the areas of Business Intelligence and Analytics, Big Data Analytics, (Spatial) Data Mining, Spatio-temporal Data Models, and Spatial Reasoning. Carlos is an Invited Lecturer in the area of Big Data and Data-oriented Systems. Both have played an important role in the development of major prototypes, working in collaboration with industry and government sectors. This book has immensely gained from their innovative research and solid experience, and will be an influential contribution to benchmarking, prototyping, and data modeling discussion in this exciting field.

Monica Wachowicz

Full Professor in Data Science
Cisco Innovation Chair in Big Data
NSERC/Cisco Industrial Research Chair in Real-time Mobility Analytics
University of New Brunswick, Canada

NOTATION

3NF – Third Normal Form

ACID – Atomicity, Consistency, Isolation, and Durability

ADM – Analytical Data Model

ADM_{BD} – Analytical Data Model in Big Data contexts

ADM_{BI} – Analytical Data Model in traditional Business Intelligence contexts

AGPL – Affero General Public License

API – Application Programming Interface

BASE – Basically Available, Soft state, and Eventual consistency

BDW – Big Data Warehouse

BDWing – Big Data Warehousing

BI – Business Intelligence

BI&A – Business Intelligence and Analytics

BSON – Binary JavaScript Object Notation

CAP – Consistency, Availability, and Partition tolerance

CDM – Columnar Data Model

CLI – Command Line Interface

CPE – Collection, Preparation, and Enrichment

CPU – Central Processing Unit

CRUD – Create, Read, Update, and Delete

DBMS – Database Management System

DDL – Data Definition Language

DDM – Dimensional Data Model

DML – Data Manipulation Language

DW – Data Warehouse

DWing – Data Warehousing

ELT – Extract, Load, and Transform

ETL – Extract, Transform, and Load

FK – Foreign Key

GFS – Google File System

GUI – Graphical User Interface

HDFS – Hadoop Distributed File System

HiveQL – Hive Query Language

HTAP – Hybrid Transaction-Analytical Processing

HTML – Hypertext Markup Language

IaaS – Infrastructure-as-a-Service

I/O – Input/Output

IoT – Internet of Things

JBOD – Just a Bunch of Disks

JDBC – Java Database Connectivity

JSON – JavaScript Object Notation

MPP – Massively Parallel Processing

NBD-PWG – NIST Big Data Public Working Group

NBDRA – NIST Big Data Reference Architecture

NIST – National Institute of Standards and Technology

NK – Natural Key

NoSQL – Not Only SQL

ODBC – Open Database Connectivity

OLAP – Online Analytical Processing

OLTP – Online Transaction Processing

ORC – Optimized Row Columnar

PK – Primary Key

RAID – Redundant Array of Independent Disks

RCFile – Record Columnar File

RDBMS – Relational Database Management System

RDD – Resilient Distributed Dataset

RDM – Relational Data Model

SCD – Slowly Changing Dimension

SF – Scale Factor

SK – Surrogate Key

SOA – Service-Oriented Architecture

SQL – Structured Query Language

SSB – Star Schema Benchmark

SSD – Solid State Drive

TPC – Transaction Processing Performance Council

TPC-DS – TPC Benchmark DS

TPC-E – TPC Benchmark E

TPC-H – TPC Benchmark H

UDF – User-Defined Function

UTC – Coordinated Universal Time

WGS84 – World Geodetic System 1984

XML – Extensible Markup Language

YARN – Yet Another Resource Negotiator

1 INTRODUCTION

Our world is generating data at unprecedented rates, mainly due to recent technological advancements in cloud computing, internet, mobile devices, and embedded sensors (Dumbill, 2013; Villars, Olofson, & Eastwood, 2011). Collecting, storing, processing, and analyzing all this data is increasingly challenging, but organizations able to do it and extract business value from it will gain significant competitive advantages. They will be able to better analyze and understand their products, stakeholders, and transactions. Big Data is frequently regarded as a buzzword for smarter and more insightful data analyzes, but arguably, it is more than that; it is about new, challenging, and more granular data sources, which require the use of advanced analytics to create or improve products, processes, and services, as well as adapting rapidly to business changes (Davenport, Barth, & Bean, 2012).

Over the last years, there has been a growing interest in Big Data (Google Trends, 2018), which is sometimes highlighted as a key asset for productivity growth, innovation, and customer relationship, benefiting business areas such as healthcare, the public sector, retail, manufacturing, and modern cities, among others (Manyika et al., 2011; M. Chen, Mao, & Liu, 2014). The definition of Big Data is ambiguous and it is difficult to determine the point at which data becomes "big" (Ward & Barker, 2013). Therefore, Big Data is frequently defined by its characteristics (e.g., its volume, variety, and velocity) and the consequent technological limitations it imposes to organizations, i.e., data that is "*too big, too fast, or too hard for existing tools to process*" (Madden, 2012, p. 4). It can be said that if Big Data is data that creates technological limitations, then it has always existed, and it always will. Currently, there is a paradigm shift in the way we collect, store, process, and analyze data. Organizations need to be aware of these technological trends and other strategies that may improve business value. Consequently, research and practice in Big Data have major relevance to ensure that organizations have rigorous proofs that emergent techniques and technologies can help them to succeed in data-driven business environments.

Big Data brings numerous research challenges, which can be divided into four main categories: i) general dilemmas, such as the lack of consensus and rigor in the definition of the concept, models, and architectures; ii) challenges related to the Big Data life cycle, from collection to analysis; iii) challenges related to security, privacy, and monitoring; iv) organizational change, such as new required skills (e.g., data scientists) or changes in workflows to accommodate the data-driven mindset. Working with Big Data requires knowledge from multiple disciplines. The term "data science" is frequently used to designate the area responsible for dealing with Big Data in all the stages of its life cycle, which relies on the scientific method (defining hypotheses and validating conclusions) and on knowledge related to areas such as machine learning, programming, and databases, to name a few. Therefore, in this book, data science is referred to as the act of extracting patterns and trends from data, using certain data-related techniques, regardless of its characteristics or challenges. These insights can then be communicated or used to create data artifacts or to optimize existing ones, improving business management and performance through data-driven decision-making (C. Costa & Santos, 2017b). In this book, the term data science is used with the meaning afore-mentioned; terms such as data mining, for example, are seen as embedded in the knowledge of data scientists and are therefore referred to as "data science techniques" (C. Costa & Santos, 2017b).

The traditional Data Warehouse (DW) is a fundamental enterprise asset, which leverages data access, analysis, and presentation in appropriate forms to support fact-based decision-making in organizations (Kimball & Ross, 2013). Recently, however, the data science community has started to question what is the role of the DW in the current era of Big Data? Which considerations for Big Data environments will lead to the redesign of the traditional DW based on relational databases? Which are the main characteristics of a Big Data Warehouse (BDW) and how it can be designed and implemented? These questions are of major relevance to understand the role of such data asset in current data-driven environments, which are dominated by volume, variety, velocity, and advanced data analytics, which impose difficulties to traditional techniques and technologies (Russom, 2014, 2016). Organizations in today's world need to understand if their current DWs are limited by the amount, structure, or velocity of data they can process, as well as to consider leveraging data science capabilities throughout their daily activities.

A BDW can be defined by its characteristics, including parallel/distributed storage and processing, real-time capabilities, scalability, elasticity, high performance, flexible storage, commodity hardware, interoperability, and support for mixed and complex analytics. BDW is a recent concept, and research on the topic is emerging but remains scarce. A critical gap identified in the literature is the lack of an integrated and validated approach for designing and implementing both the logical layer (data models, data flows, and interoperability between components) and the physical layer (technological infrastructure). The divergence regarding the concept of BDW is alarming. Moreover, a prescriptive approach in which models, methods, and instantiations are tightly coupled is still needed (thus providing a cohesive way of building BDWs).

The current trend in BDW design and implementation mainly consists in finding the best technology to meet Big Data demands (use case driven approach), instead of following a data modeling approach (data-driven approach) (Clegg, 2015). Although there are already some best practices, non-structured guidelines, and implementations in specific contexts, they do not cover many of the characteristics of a BDW identified in the literature. Since works related to the BDW concept are multidisciplinary, certain approaches focus on general guidelines and best practices, while others focus on the technological advancements in storage and analytics. However, as previously mentioned, there is no integrated approach focusing on both the logical and the physical layers to implement the characteristics of a BDW with adequate evaluation (e.g., benchmarking, prototypes, and data modeling discussion), which could provide a general-purpose approach, and prescribe models and methods for researchers and practitioners.

1.1. OBJECTIVES OF THIS BOOK

The main problem identified in the literature is that there is a significant gap between *this is what a BDW should be* and *this is how it must be designed and implemented*, leading to a use case driven approach primarily concerned with finding the best technology to meet demands. The proposal of a prescriptive approach to design and implement BDWs contributes to the development of new initiatives in a rigorously justified manner, wherein models (representations of logical and infrastructural components), methods (structured practices), and instantiations (prototypes or implemented systems) are tightly coupled and grounded on evaluated practices. Such contribution aims to enrich data-driven approaches in Big Data environments, in which the models and methods are so general that the context of the instantiations becomes as irrelevant as possible, similarly to what usually happens in traditional DWs. Big Data brings major changes in the way one is used to build traditional DWs, including different techniques and technologies. However, this book assumes that such changes do not imply discarding the relevance of data models and methods in favor of a use case driven approach. Consequently, the main goal of this book is presenting a general-purpose approach for designing and implementing BDWs, wherein models and methods are adequately integrated and validated, while providing practitioners a broad overview of Big Data technologies that can be used in these analytical contexts, as well as examples of real-world applications.

The approach shown in this book is the result of an extensive design science research process (Hevner, March, Park, & Ram, 2004; Peffers, Tuunanen, Rothenberger, & Chatterjee, 2007), and provides a set of models and methods to guide practitioners working in this area. Also, it aims to foster future research related to BDWs, by inviting researchers and practitioners to further evaluate it in several implementation contexts. This work is not focused on "lift and shift" strategies, therefore, the coexistence of the traditional DW with Big Data technologies is not considered here. Consequently, we foresee the use of the proposed approach in the following scenarios: i) the organization does not currently have a traditional DW and wants to implement a modern data asset, namely a BDW; ii) the organization has a traditional DW and wants to replace it ("rip and replace" strategy); iii) or, finally, the organization relies on a use case driven approach, maintaining a complex and not interoperable federation of different technologies, and wants a data-driven approach with high interoperability between components and well-defined methods, data models, and data flows.

BDWs have major relevance for the scientific community and practitioners in the area of data engineering and data science. Developing an approach in which models, methods, and instantiations are tightly coupled and scientifically evaluated will provide a structured guide to DW practitioners and promote future research on the concept of BDW. Taking the former into consideration, the objectives of this book are the following:

1. Present the main Big Data concepts, techniques, and technologies (with practical examples when applicable), such as:

    a) Big Data definition, relevance, and challenges;

    b) Design guidelines/techniques such as the National Institute of Standards and Technology (NIST) Big Data Reference Architecture (NBDRA) (NBD-PWG, 2015), the Big Data Processing Flow (Krishnan, 2013), the Data Highway Concept (Kimball & Ross, 2013), and the Lambda Architecture (Marz & Warren, 2015);

    c) Big Data technologies like Hadoop and its ecosystem, distributed Structured Query Language (SQL) engines (e.g., SQL-on-Hadoop systems), and Not Only SQL (NoSQL) or NewSQL databases.

2. Disseminate models and methods for BDW design and implementation:

    a) A model of the logical components and their interoperability, representing how data flows through the different components and detailing how they interchange data according to the proposed data modeling method;

    b) A method for collecting, preparing, and enriching data flowing to the BDW, including structured, semi-structured, and unstructured data. Data science techniques (e.g., data mining and text mining) are taken into consideration, in order to give structure to the data and deliver predictive capabilities. This method also includes concerns regarding batch data and streaming data (low latency and high frequency);

    c) A technological infrastructure model, representing how the Big Data technologies can be used, organized, and deployed in a shared-nothing architecture;

    d) A data modeling method that can accommodate all types of data regardless of their structure and subject. Obviously, unstructured data does not fit into predefined data models and, therefore, in this case, data mining and text mining techniques are used to extract value from data, giving structure to the relevant findings, and storing those in the BDW. Consequently, the BDW will not only have historical data, but also real-time data and predictive capabilities.

3. Elucidate the intended audience regarding the use of the aforementioned models and methods using demonstration cases:

    a) Demonstrate the proposed data modeling method when applied to different real-world problems (e.g., retail, manufacturing, finance, sensor-based analysis, and digital media). This objective is focused on creating a set of BDW data models

© FCA

and examples of data modeling guidelines available, which practitioners can take into consideration when building their own applications;

b) Present the design and implementation details regarding batch and streaming data Collection, Preparation, and Enrichment (CPE) processes. Batch processes do not aim for low latency and high frequency, unlike streaming processes, in which each event should be loaded into the BDW with a latency between milliseconds and a few seconds. This demonstration case also considers how several data science techniques (e.g., data mining and text mining) can be efficiently included in batch and streaming data CPE processes, as the approach aims to support the design and implementation of both descriptive and predictive BDWs;

c) Discuss the benchmark of various workloads and scenarios, including different Scale Factors (SFs) and dimensions size for batch data, use of data partitioning, use of nested attributes, drill across, window and analytics functions, concurrent workloads, and stream processing. This will allow for the evaluation of how a BDW created using the proposed approach handles large scans needed for *ad hoc* analysis, reporting, and data visualization, compared to a traditional dimensional DW, as well as how it handles streaming scenarios, concurrent workloads, and semi-structured analytics (e.g., analysis using nested arrays, key-value pairs, or geometry objects);

d) Disseminate the suitability of the proposed approach for solving real-world BDWing problems, by presenting the implementation of a BDWing system for smart cities that follows the proposed models and methods. In this case, the SusCity research project is used. This focuses on the development and integration of new tools and services to improve the efficiency of urban resources, reducing the environmental impact and promoting economic development and reliability (SusCity, 2016). The main goal is to advance the science of urban systems modeling and the data representation supported by the collection and processing of Big Data. This allows the creation of new services to explore economic opportunities and the sustainability of urban systems. The SusCity project has a testbed in Lisbon that includes several data sources (e.g., sensors, census, buildings characteristics, and geolocation data related to mobility), generating data at different velocities (e.g., batch and streaming), with significant volume. Moreover, data science techniques, such as data mining (e.g., clustering and time series forecasting) and data visualization, are crucial to create new services to improve urban systems. Therefore, this research project is used to instantiate the approach, discussing and evaluating the proposed models and methods for collecting, storing, processing, and analyzing Big Data, thus proving the suitability of the approach to solve real-world problems.

1.2. INTENDED AUDIENCE

This book is intended for DWing and Big Data practitioners, teachers, researchers, and students focusing on the design, development, and implementation of analytical applications supported by a DW. As prerequisite knowledge, the audience should be familiar with DWs modeling strategies (e.g., dimensional modeling and third normal form "3NF" DWs).

1.3. BOOK STRUCTURE

After this introductory chapter, this book is structured as follows: Chapter 2 presents the relevance, definition, and challenges in Big Data, and the techniques and technologies for designing and implementing Big Data solutions, including relevant Big Data architectures. Chapter 3 describes data systems oriented towards transactional workloads in Big Data, such as NoSQL databases, while Chapter 4 is dedicated to analytical data systems, such as Hive and Druid. Chapter 5 describes an approach for the design and implementation of BDWs, presenting the general models and methods, the suggested technological infrastructure, and the data modeling method. Chapter 6 presents several BDW data models and data modeling considerations that practitioners can follow to implement BDW applications. Chapter 7 shows several workloads for collecting, preparing, and enriching data, facilitating the understanding of the adequate mechanisms for handling data throughout these three stages. Chapter 8 evaluates the performance of BDWs by benchmarking various design patterns introduced in Chapter 5 and in Chapter 6. Chapter 9 presents a real-world BDW application in the context of smart cities, namely discussing the implementation of the SusCity project, from data collection to data visualization. Chapter 10 concludes with a synopsis of the book and the contributions to the state of the art in this field.

2 BIG DATA CONCEPTS, TECHNIQUES, AND TECHNOLOGIES

INTRODUCTION

We live in a world where data is constantly produced and consumed, so understanding the value that can be extracted from it is a priority. Organizations need to understand and analyze relevant data flows, join data analytics with product/process development, and move it closer to the core business (Davenport et al., 2012). This chapter shows the relevance Big Data has in today's world, various attempts to define it, the related challenges, and several techniques and technologies to efficiently design and implement Big Data solutions (C. Costa & Santos, 2017a).

2.1. BIG DATA RELEVANCE

Over the last years, the interest in Big Data has increased considerably (Google Trends, 2018), particularly after 2012, as can be seen in Figure 2-1. In a McKinsey Global Institute's report, Manyika et al. (2011) argue that Big Data will become fundamental among organizations for productivity growth, innovation, and customer relationship, highlighting its relevance in healthcare, public sector, retail, manufacturing, and personal-location contexts, stating that value can be generated in each one of them. Nowadays, data has a strong presence in the daily activities of almost every industry, alongside labor and capital, as Manyika et al. (2011) also demonstrated by estimating that in 2009 almost all economic sectors in the United States had nearly 200TB of stored data per organization with more than 1000 employees. Other statistics similarly show that the amount of data available in today's world is growing exponentially (Chandarana & Vijayalakshmi, 2014).

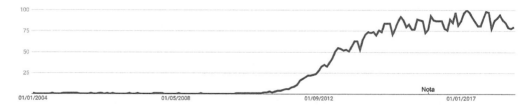

Figure 2-1. Increased interest in Big Data. Reprinted from Google Trends (2018).

Nevertheless, as human beings tend to resist to changes, there are still those who ask themselves: "Why Big Data? Why Now?" (Krishnan, 2013). According to Krishnan (2013), the concept of Big Data is about leveraging access to a vast volume of data, which can help retrieving value for organizations, with minimal human intervention, due to the advancements made in data processing technologies. The author also claims that Big Data always existed in many industries, but the emergence of autonomous, fast, flexible, and scalable processes created a new paradigm shift, often resulting in a cost reduction when compared to traditional data processing approaches.

Organizations find themselves facing this new data-driven way to conduct business, and a paradigm shift in their infrastructure and way of thinking is, understandably, something that requires serious consideration. However, they need to foresee the value that Big Data can bring to their business (Manyika et al., 2011):

- The use of Big Data can make information more transparent and usable across the organization;

- Business performance can be increased with more accurate and detailed facts, turned possible by collecting and processing more transactional data;

- Better management decisions can be made through data analysis;

- The use of Big Data has the ability to refine and reinvent products and services.

Even so, the evidence that using Big Data intelligently will improve business performance can still be questioned, as McAfee, Brynjolfsson, Davenport, Patil, and Barton (2012) highlight by discussing the inadequacy of the business press to demonstrate the real value of being data-driven and testing the hypothesis that data-driven organizations are better performers than traditional ones. The authors interviewed executives in 330 organizations and also gathered performance data about their respective organizations. McAfee et al. (2012) came to an interesting conclusion: organizations that see themselves as data-driven achieved better performance regarding financial and operational goals. The authors highlight increased productivity and profitability in top organizations that used data-driven decision-making, even when taking into consideration other factors like labor and capital, for example. The results shown by McAfee et al. (2012) rigorously corroborate the current trend for Big Data value within organizations. The use of Big Data will become inevitable for competitive advantages across most industries, from electronic and information industries to finance, insurance, or government. Big Data can leverage increasing productivity and better customer relationship (Manyika et al., 2011), and can potentially be used in several business areas to generate significant value for organizations (Chandarana & Vijayalakshmi, 2014; Manyika et al., 2011; Villars et al., 2011), as Table 2.1 shows.

According to Brown, Chui, and Manyika (2011), other business areas are worth mentioning, such as finance, insurance, and real estate. The authors present an approach that analyzes several business areas by the ease-of-capture Big Data and its potential to generate value. The apparent trend is for organizations to perceive value in data-driven decision-making and start collecting more data, contributing to the continuous growth in data volume. Big Data will have a significant impact in value creation and competitive advantage for organizations, such as new ways to interact with customers or to develop products, services, and strategies, consequently raising profitability. Another area where the concept of Big Data is of major relevance is the Internet of Things (IoT), seen as a network of sensors embedded into several devices (e.g., appliances, smartphones, cars), which is a significant source of Big Data, bringing many business environments (e.g., cities) into the era of Big Data (M. Chen et al., 2014).

As noted above, Big Data brings competitive advantages to organizations, but there are particular characteristics that define it, although most of the times they are unquantifiable (Ward & Barker, 2013), as will be discussed in the next section. Big Data creates a new paradigm shift in the way we collect, store, process, and analyze data, but organizations can be data-driven and explore the potential of data from innumerous sources without dealing with Big Data techniques and technologies. In that case, they are just dealing

© FCA

with new data, data that was not previously processed within the organization, but does not bring severe difficulties for the capabilities of traditional techniques and technologies.

In the next section, the definition of Big Data will be discussed following the perspective of different authors.

Table 2.1. Big Data applied in several business areas.

Business Area	Examples of Application
Healthcare	• Personalize medication and understand causes of diseases, using techniques to extract value from vast amounts of data on medical history, medication, and drug manufacturing, for example; • Analyze genetic variations and the effectiveness of potential treatments, using vast amounts of data; • Other Big Data sources can include nutrition and training data or even more unstructured data like medical images.
Environment	• Find a correlation between the measured values and the implications for the environment through the collection of data from multiple sensors (e.g., air and water quality, meteorology, and gas emissions).
Public sector	• Use Big Data to prevent tax fraud and errors; • Customize actions by segmenting population; • Create more transparency through data availability.
Retail	• Event forecasting and customer segmentation, creating personalized products or services; • Location based marketing, sentiment analysis, and cross-selling; • Logistics optimization.
Manufacturing	• Demand forecasting for supply planning; • Use of sensors in manufactured products to offer proactive maintenance.

2.2. BIG DATA CHARACTERISTICS

Technological advancements have opened the way for the generation of unprecedented amounts of data each day, at ever-increasing rates. In 2011, around 1.8 zettabytes of data were produced every other day, that is more than it was produced since the beginning of civilization until 2003 (M. Chen et al., 2014). As a result, storage capacity must increase, and new ways of dealing with such amounts of data need to be developed. At this point, the relevance of Big Data, the way it is changing how organizations operate, and the opportunities that data-driven approaches bring are clear, but what does Big Data actually mean?

First of all, there is no widely accepted threshold for classifying data as Big Data. Ward and Barker (2013), in an attempt to clearly define Big Data, present several notorious definitions among the community, highlighting that Big Data is predominantly and "anecdotally" associated with data storage and data analysis, terms dating back to distant times, and argue that the adjective "big" implies significance, complexity, and challenge, but it also makes it difficult to quantitatively define Big Data. Ward and Barker (2013) present several definitions: some define Big Data by its characteristics, others based on the augmentation of traditional data with more unstructured data sources, and some using quantitative thresholds. They also present definitions which rely on the inadequacy of traditional technologies to deal with this new type of data, presenting various perspectives from the industry, including Gartner, Oracle, Intel, Microsoft, and IBM, for example. Ward and Barker (2013) note that all definitions include at least one of the following aspects: size, complexity, or techniques and technologies to process large and complex datasets.

For his part, Dumbill (2013, p. 1) provides a definition based on technological constraints: "*Big Data is data that exceeds the processing capacity of conventional database systems. The data is too big, moves too fast, or does not fit the strictures of your database architectures. To gain value from this data, you must choose an alternative way to process it*". M. Chen et al. (2014) corroborate this definition by focusing on the fact that traditional software and hardware cannot recognize, collect, manage, or process this new type of data in reasonable time. Krishnan (2013) also agrees with these perspectives, defining Big Data by its complexity, creation speed, and several levels of ambiguity, whose processing is inadequate for traditional methods, algorithms, and technologies. Although Ward and Barker (2013) are slightly critical of both the lack of quantification in Big Data's definition and the use of data storage and analysis terms when defining the Big Data concept, they conclude that the same includes storage and analysis of large and complex datasets, using a set of novel techniques.

The actual origins of the concept of Big Data are relatively unknown, and its definition evolved rapidly. Gandomi and Haider (2015) state that size is the characteristic that stands out the most, but other features became more common to define Big Data. In 2001, Doug Laney, from Gartner, presented the 3Vs model (Figure 2-2) to characterize Big Data by its volume, variety, and velocity (Laney, 2001). IBM and Microsoft based their definitions of Big Data on this model for at least ten more years (M. Chen et al., 2014).

According to Gandomi and Haider (2015), volume is a characteristic which indicates the magnitude of data, noting that it is frequently reported between Terabytes and Petabytes, citing the survey of Schroeck, Shockley, Janet, Romero-Morales, and Tufano (2012), wherein just over half of the respondents consider datasets bigger than 1TB to be Big Data. However, the authors argue that data size is relative and varies according to the periodicity and the type of data. It is impractical to define a specific threshold for Big Data volume, as different types of data require different technologies to deal with it (e.g., tabular data and video data), as Gandomi and Haider (2015) exemplify. The volume in the 3Vs model characterizes the amount of data that is continuously generated (Krishnan, 2013),

© FCA

and the main cause for the ever-increasing volume is the fact that we currently store all our interactions with the majority of services available in our world (Zikopoulos & Eaton, 2011).

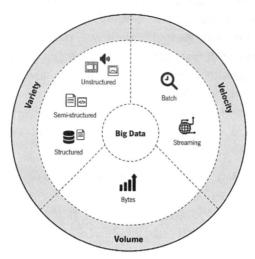

Figure 2-2. The 3Vs model. Adapted from Zikopoulos & Eaton (2011).

Regarding variety, Big Data can be classified as structured (e.g., transactional data, spreadsheets, and relational databases), semi-structured (e.g., Web server logs, Extensible Markup Language – XML, and JavaScript Object Notation – JSON), and unstructured (e.g., social media posts, audio, video, and images) (Chandarana & Vijayalakshmi, 2014; Gandomi & Haider, 2015). Traditional technologies can have significant difficulties storing and processing Big Data, such as content from Web pages, click-stream data, search indexes, social media posts, emails, documents, and sensor data. Most of this data does not fit well in traditional databases, hence, there must be a paradigm shift in the way organizations perform analyzes to accommodate raw structured, semi-structured, and unstructured data, in order to take advantage of the value in Big Data (Zikopoulos & Eaton, 2011).

The final characteristic covered by the 3Vs model is velocity, which refers either to the rate at which data is generated or to the required speed of analysis and decision support (Gandomi & Haider, 2015). Data can be generated at different rates, ranging from batch to real-time (streaming) (Chandarana & Vijayalakshmi, 2014; Zikopoulos & Eaton, 2011). It is preferable to apply the definition of velocity to data in motion, instead of the rate at which data is collected, stored, and retrieved from storage. Continuous data streams can create competitive advantages in contexts where the identification of trends must occur in short periods of time, as in financial markets, for example (Zikopoulos & Eaton, 2011).

Over time, two additional characteristics associated with Big Data were identified: value and veracity. Value represents the expected result of processing and analyzing Big Data

(Chandarana & Vijayalakshmi, 2014), which is usually low in raw data, since this is mainly extracted with an adequate analysis (Gandomi & Haider, 2015). According to Chandarana and Vijayalakshmi (2014), value can be obtained by integrating different data types to improve efficiency and gain competitive advantages. On the other hand, veracity draws attention to possible imprecise data, as the analysis can be based on datasets with varying degrees of precision, authenticity, and trustworthiness (Chandarana & Vijayalakshmi, 2014). Gandomi and Haider (2015) corroborate this definition, highlighting the unreliability of certain data sources (e.g., customer sentiments extracted from social media), although recognizing that they can be valuable when adequate techniques and technologies are used.

Other characteristics, which according to the literature often go unnoticed, are the variability and complexity, introduced by SAS (Gandomi & Haider, 2015). Variability is related to the different rates at which data flows, according to different peaks and inconsistent data velocity. Complexity highlights the challenges of dealing with data from multiple sources, namely connecting, matching, cleaning, and transforming them. Besides the former, Krishnan (2013) also proposes three other characteristics: ambiguity, concerning the lack of appropriate metadata, resulting from the combination of volume and variety; viscosity, when the volume and velocity of data cause resistance in data flows; virality, which measures the time of data propagation among peers in a network. Figure 2-3 presents a summary of all these characteristics identified in the literature.

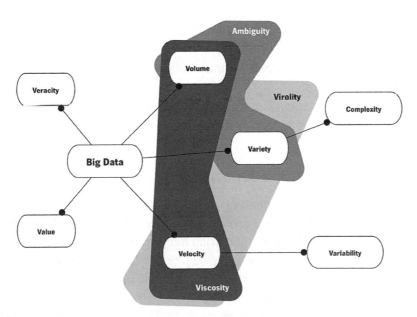

Figure 2-3. Main Big Data characteristics identified in the literature. Adapted from C. Costa & Santos (2017a).

© FCA

At this point, it seems that trying to quantify any of these characteristics is an impossible task. Big Data remains an abstract concept (M. Chen et al., 2014). It can be the combination of different characteristics or it can be mostly the presence of one of them, but it is data that requires changes to the techniques and technologies needed to process it. It may be a database or a DW that cannot be scaled accordingly on a shared-everything architecture (Krishnan, 2013), or data mining tasks that cannot be accomplished without parallel computing. Again, Big Data is *"too big, too fast, or too hard for existing tools to process"* (Madden, 2012, p. 4). Defining Big Data by the inadequacy of traditional technologies is relatively dangerous, as advancements are constantly being made (e.g. quantum computing); furthermore such definition implies that Big Data has always existed and will continue to exist (Ward & Barker, 2013). The current definitions of Big Data are relatively dependent on the techniques and technologies for collecting, storing, processing, and analyzing it. These techniques and technologies will evolve over time and we need to learn to adapt ourselves to those changes. Analyzing new technological trends that may benefit business and reconsider new strategies related to data will always be important. Currently, a new paradigm shift is happening and it does not need to impact all organizations. Nevertheless, scientific progress in this field will continue providing to organizations evaluated and efficient techniques and technologies for addressing data-driven environments. The state of the art regarding Big Data techniques and technologies will be presented later in this book. The next section presents the challenges regarding Big Data.

2.3. BIG DATA CHALLENGES

This section presents several challenges associated with handling Big Data, including general dilemmas, challenges in the Big Data life cycle, issues concerning security, privacy, and monitoring, as well as the changes organizations may need to undergo. These challenges also serve to identify relevant research topics across various fields.

2.3.1. Big Data General Dilemmas

General dilemmas may include challenges such as the lack of consensus and rigor in Big Data's definition, models, and architectures. For example, M. Chen et al. (2014) claim that the concept of Big Data often has more to do with commercial speculation than with scientific research. The authors also mention the lack of standardization in Big Data, such as data quality evaluation and benchmarking. In fact, the lack of standard benchmarks to compare different technologies is seriously aggravated by the constant technological evolution of Big Data environments (Baru, Bhandarkar, Nambiar, Poess, & Rabl, 2013).

How to take full advantage of Big Data in areas such as scientific research, engineering, medicine, finance, education, government, retail, transportation, or telecommunications remains an open question (M. Chen et al., 2014). Discussions about how to select the most appropriate data from several sources or how to estimate their value are major issues (Chandarana & Vijayalakshmi, 2014). Another issue regularly discussed is how Big Data helps representing the population better than a small dataset does (Fisher, DeLine, Czerwinski, & Drucker, 2012). The answer obviously depends on the context, but the authors make the important point that one should not assume that more data is always better.

2.3.2. Challenges in the Big Data Life Cycle

The following challenges are related to technical difficulties when performing tasks such as Big Data collection, integration, cleansing, transformation, storage, processing, analysis, and governance:

- The need to rethink storage devices, architectures, mechanisms, and networks, in order to achieve more efficient input/output (I/O), data accessibility, and data transmission (C. L. P. Chen & Zhang, 2014);

- Scalability becomes crucial to store and analyze data. Handling increasing amounts of data requires redesigning databases and algorithms to extract value from it (Hashem et al., 2015). Distributed/parallel computing becomes crucial to deal with Big Data, assuring availability, cost efficiency, and elasticity (M. Chen et al., 2014);

- Guaranteeing data quality and adding value through data preparation becomes challenging in Big Data environments (C. L. P. Chen & Zhang, 2014). Different data sources may have different data quality problems (Hashem et al., 2015). These problems and vast amounts of redundancy can also make data integration more difficult (M. Chen et al., 2014). The heterogeneity resulting from multiple sources augments these challenges, since traditional data analysis techniques expect homogeneous data (Jagadish et al., 2014). Heterogeneity has implications over data integration (Cuzzocrea, Song, & Davis, 2011) and the analysis of Big Data, as the unstructured nature of data sources presents several challenges regarding transformations for supporting adequate analytical tasks;

- Visualizing Big Data requires rethinking traditional approaches due to the volume of data, thus combining appearance and functionality is crucial (C. L. P. Chen & Zhang, 2014). Advanced data visualizations are needed to extract value from Big Data (Russom, 2011), providing the capacity to scale to thousands or millions of data points, handle multiple data types, and enhancing ease of use to accomodate multiple users. Krishnan (2013) argues that manipulating Big Data is challenging due to the difficulty of executing drilldowns or rollups. In these visualizations, data from multiple sources is typically integrated into a single picture. The author indicates that technological evolutions would be made to address the challenge of manipulating Big Data interactively, as discussed later in this document;

- Searching, mining, and analyzing Big Data is a challenging and relevant research trend, including Big Data searching algorithms, recommendation systems, real-time Big Data mining, image mining, text mining, among others (M. Chen et al., 2014). As Gandomi and Haider (2015) claim, size is frequently the main concern in Big Data, but the unstructured nature of certain data also deserves attention (e.g., text, audio, and video) and imposes significant challenges on these tasks;

- Big Data governance faces challenges regarding control and authority over massive amounts of data from different sources (Hashem et al., 2015). Managing such heterogeneous environment to plan access policies and ensure traceability can quickly become almost impossible without adequate governance tools.

Organizations face several challenges in the Big Data life cycle. New business problems require technological innovations concerning the way data flows throughout the organization. Overcoming these challenges will depend on the organization's maturity, since legacy applications and the use of incompatible formats can impose several limitations for an adequate integration and extraction of value from Big Data. Collecting data, namely gaining access to it, may also be a challenge, as integrating data from multiple sources, including external ones, raises questions about other actor's reasons for sharing it free of charge (Manyika et al., 2011).

M. Chen et al. (2014) state that the efficiency in data flows is a key factor for ensuring an adequate processing of Big Data. The authors also highlight the challenge of building effective real-time computing models and online applications to analyze Big Data. Other challenges related to processing Big Data may include the reutilization and reorganization of data, which become laborious at large scales. Due to its characteristics, Big Data requires a paradigm shift in databases and analytical technologies, since handling Big Data throughout its life cycle can potentially create severe bottlenecks in networks, storage devices, and relational databases. Technology is evolving to execute these stages across distributed environments, thus increasing dependency on high storage capacity and processing power.

Relational databases are evolving and adapting to the above trends, increasing query performance and including capabilities to deal with more data variety (Davenport et al., 2012). Combining the benefits of a Relational Database Management System (RDBMS) with the new trends in data systems for handling Big Data is a current research trend, as well as query optimization mechanisms for Big Data technologies (Cuzzocrea et al., 2011). Furthermore, advancements are constantly being made in scalable storage and algorithms. Ji, Li, Qiu, Awada, and Li (2012) argue that processing queries in Big Data may take significant time, as it is challenging to sequentially iterate through the whole dataset in a short amount of time. Consequently, the authors highlight the importance of designing indexes and implementing adequate preprocessing technologies. Hashem et al. (2015) identify the need to study adequate models to store and retrieve data as a crucial factor to successfully implement Big Data solutions. Models and algorithms for scalable data analysis also remain an open research issue, as well as the integration and analysis of data arriving continuously from streams. Mining data streams has been identified as an emergent research topic in Big Data analytics (H. Chen, Chiang, & Storey, 2012).

2.3.3. Big Data in Secure, Private, and Monitored Environments

Nowadays, keeping data secure and private is one of the most important tasks for organizations (M. Chen et al., 2014; Jagadish et al., 2014). Users want to be sure that their data will not be leaked or stolen (Chandarana & Vijayalakshmi, 2014). Sagiroglu and Sinanc (2013), citing a survey from (Intel IT Center, 2012), claim that security and privacy are frequently mentioned among the Big Data concerns of IT managers. It is thus important to plan a Big Data driven security model so that organizations can accurately specify risks and prevent illegal activities or cyber threats. Among the relevant considerations mentioned by the literature are authentication, authorization, network traffic analysis, data protection laws, and data mining related to security. M. Chen et al. (2014) also discuss the potential for Big Data applications related to security concerns.

Due to the characteristics of Big Data, more security and privacy risks arise, and traditional data protection methods must be redesigned. M. Chen et al. (2014) argue that Big Data applications face multiple challenges related to security, privacy, and monitoring: protection of personal privacy not only during data collection, but also in subsequent storage and flows; Big Data quality and its influence on the appropriate and secure use of data; the performance of security mechanisms like encryption is largely influenced by the scale and variety of data; and other aspects related to secure communications, administration, and monitoring in environments with multiple users and services. Another relevant challenge, as highlighted by Hashem et al. (2015), is ensuring Big Data integrity, i.e., that data can be modified only by the owner or authorized third parties.

Policies related to data handling are more relevant nowadays, when there is a significant amount of sensitive data about individuals, such as that concerning their health and finances (Manyika et al., 2011). Legal issues are being raised by how easy it is to copy, integrate, and recurrently use data by different people. Intellectual property, data ownership, and responsibility regarding inaccurate data deserve proper attention from policy-makers (Manyika et al., 2011). Legal and regulatory issues also deserve attention (Ji et al., 2012) in several aspects, like analyzing whether current laws and regulations adequately protect data about individuals (Hashem et al., 2015). Even the constant tracking of employees within an organization can raise issues regarding adequate work policies (Michael & Miller, 2013).

Aside from the above issues, Brown et al. (2011) address the implications of having data widely and transparently available. Organizations that rely on costly proprietary data to leverage their competitive advantages will face challenges due to the promises of more accessible Big Data sources, as they become widely available in some contexts. The authors also discuss the inherent difficulties that organizations face when sharing data across departments and forming a coherent view of the organization, which is additionally aggravated with Big Data. Organizations need to integrate data from multiple sources and promote collaboration, not only among departments, but also among suppliers and customers (Brown et al., 2011).

Ensuring privacy is both a technical and a sociological problem, as Jagadish et al. (2014) argue. The inadequate availability of location-based data makes it possible to infer a person's residence, office location, and identity, for example. Moreover, many other data sources can contain personal identifiers, and even if they do not, when the data is rich enough, reasonable inferences can be drawn from it (Wigan & Clarke, 2013). Currently, we tend to share more data online, most of the times without knowing the implications. Data ownership is another relevant topic, as organizations want to share or sell data, due to its value, without losing its control (Jagadish et al., 2014). Data ownership is often discussed regarding social media websites, since the users' data is not owned by the organizations although they are the ones storing it (Chandarana & Vijayalakshmi, 2014). As Wigan and Clarke (2013) note, these organizations tend to assume that they hold the rights of the data, and sometimes the current legislation benefits them, allowing them to keep the data, even when users explicitly ask for it to be deleted.

Big Data security, privacy, and monitoring in cloud environments is also a relevant topic for discussion. Organizations frequently recognize that using Big Data technologies in cloud environments helps reducing their IT costs (Ji et al., 2012), although raising concerns about Big Data storage and processing infrastructures. Therefore, as Ji et al. (2012) claim, one of the challenges lies in guaranteeing adequate monitoring and security without exposing users' data when processing it.

2.3.4. Organizational Change

Big Data may sound appealing to most organizations, but frequently leaders lack knowledge concerning its value and how to extract it (Manyika et al., 2011). Occasionally, the lack of knowledge on how to use analytics is mentioned as the main obstacle to become more data-driven (LaValle, Lesser, Shockley, Hopkins, & Kruschwitz, 2011). Within several business areas, organizations need to monitor trends and gain advantages compared to their competitors, but as Manyika et al. (2011) discuss, many lack the talent, the rigorous workflows structure, and the incentives for implementing Big Data initiatives to support their decision-making. Leaders and policy-makers must understand how Big Data can create value, as well as critically think about IT capabilities, data strategies, analytical talent, and data-driven approaches. This paradigm shift in organizations requires them to place analytics in the core business and operational functions (Davenport et al., 2012), changing business processes, and delivering insights related to customers, products, services, and other transactions. McAfee et al. (2012) describe five challenges that organizations will face in management, caused by Big Data initiatives:

- Ensure adequate leadership for a Big Data project;

- Find suitable data scientists (Provost & Fawcett, 2013), computer scientists, and other professionals to deal with Big Data, design experiments, and overcome business challenges;

- Understand and adequately use Big Data technology;

- Pair problem-solvers with the right data for decision-making;

- Change organizational culture and rethink how data-driven the organization really is.

Big Data initiatives require a multidisciplinary approach and collaboration to deliver useful results that will be understandable by the organization (Jagadish et al., 2014), but to accomplish this, challenging organizational changes must occur.

2.4. TECHNIQUES FOR BIG DATA SOLUTIONS

As previously mentioned, the concept of Big Data is often used to sell something (Fan & Bifet, 2013), denoting a lack of common understanding about the concept itself, and opening the way for the proposal of an almost infinite set of techniques and technologies. Unfortunately, this raises significant challenges for understanding, adopting, or designing techniques to work with Big Data, since they are tightly coupled with a specific technology. The opposite problem can also occur, since most of the times, in conceptual models, it is not clear which technology should be employed in a given component of the model. This is mainly due to Big Data's variety, but even discarding unstructured data (e.g., text, video, image, and audio), it seems that almost everyone is trying to sell their solutions, mainly following a use case driven approach, without concerns regarding a common way to design and implement solutions.

The age of Big Data can generate significant controversy. For example, Fan and Bifet (2013), citing Boyd and Crawford (2012), claim that it is not necessary to distinguish Big Data analytics from data analytics, as data volume will continue to grow and never decrease. The transition from traditional techniques and technologies is a radical paradigm shift, which can lead to abandoning shared-everything architectures (Krishnan, 2013), RDBMSs, common Extract, Transform, and Load (ETL) mechanisms, or SQL, for example. The ambiguity in Big Data's definition, the lack of formal and recognized techniques, and the vast set of available technologies do not contribute to a widespread adoption of Big Data. It can be argued that the Big Data analytics area needs approaches like the widely accepted work from Kimball and Ross (2013) that focuses on how to store and analyze data in a relational DW. Guaranteeing that businesses do not refrain from progress due to uncertainty or lack of resources is of major relevance.

Regarding the question of when it would be an appropriate moment for an organization to rethink the techniques and technologies that it normally employs, a survey from Russom (2011) suggests that most organizations tend to replace traditional platforms when:

- Massive performance and scalability are required, such as the need to scale to Big Data contexts with a large volume of data, speed up data collection and queries, or ensure concurrent workloads;

- Business users need advanced analytics (e.g., data mining, statistical analysis, text analytics, and ad hoc SQL queries), and the current platform is Online Analytical Processing (OLAP) only;

- Organizations need self-service and rich visualization tools for end users;

- The platform lacks modern capabilities, such as support for a Service-Oriented Architecture (SOA), cloud infrastructures, or in-memory processing.

This section aims to present several techniques to understand and deal with Big Data throughout its life cycle, from collection to analysis, including storage and mining. These techniques mainly represent a collection of guidelines to help designing Big Data solutions, their components, the relationship between them, and some necessary changes in traditional approaches for dealing with data. According to C. L. P. Chen and Zhang (2014), citing Marz and Warren (2012) and Garber (2012), a Big Data solution generally contemplates the following principles:

- Present high-level architectures, addressing the distinct role of specific technologies;

- Include a variety of data science tasks, such as data mining, statistical analysis, machine learning, real-time visualization, and in-memory analysis;

- Combine the benefits of different tools for different tasks;

- Bring analysis closer to the data, in order to avoid moving data;

- Distribute processing and storage across different nodes in a cluster;

- Ensure coordination between data and processing nodes to improve scalability, efficiency, and fault-tolerance.

2.4.1. Big Data Life Cycle and Requirements

There are several aspects of the life cycle of Big Data, which significantly differ from traditional environments and thus need to be considered. Dealing with Big Data requires new approaches, which are discussed in this subsection.

2.4.1.1. General Steps to Process and Analyze Big Data

According to a survey conducted on analysts at Microsoft (Fisher et al., 2012), Big Data analytics tasks can be divided into five steps: acquiring data; choosing the architecture based on cost and performance; shaping the data according to the architecture; writing and editing code; and reflecting and iterating on the results. Processing Big Data for analysis differs from processing traditional transactional data. As Krishnan (2013) claims, in traditional environments, data is explored, a model is designed, and a database structure is created. However, in Big Data environments, data is first collected and loaded into a certain storage system, a metadata layer is applied, and then a structure is created. There is no need to start by transforming data to properly fit a relational model, since transformations only occur after having everything stored within efficient storage systems. This represents a shift from a traditional ETL approach to an Extract, Load and Transform (ELT) approach. Figure 2-4 illustrates the Big Data Processing Flow according to Krishnan (2013).

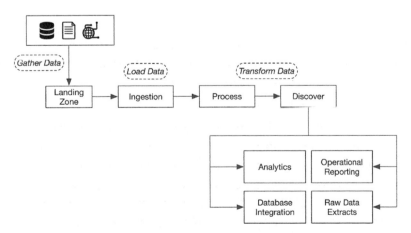

Figure 2-4. Big Data Processing Flow. Adapted from Krishnan (2013) and C. Costa & Santos (2017a).

The Big Data Processing Flow starts by gathering data from multiple sources, such as Online Transaction Processing (OLTP) systems, multiple files, sensors, and the Web. This data is then stored in a landing zone capable of handling the volume, variety, and velocity of the data, which is typically a distributed file system. Data transformations must occur on data stored in the landing zone, fulfilling the requirements of efficiency and scalability, and the subsequent results can then be integrated into analytical tasks, operational reporting, databases, or raw data extracts. Regarding this process, Kimball and Ross (2013) mention relevant best practices for the Big Data life cycle:

- Plan a "data highway" with multiple caches – raw source (immediate), real-time cache (seconds), business activity cache (minutes), top line cache (24 hours), and DW or long time series cache (daily, periodic, and annual). Data will flow through these different caches, according to the business needs;

- Use Big Data analytics to enrich data before moving it to the next cache. For example, produce numeric sentiments from mining unstructured tweets. The opposite is also true, so that earlier caches can benefit from the less granular ones. Kimball and Ross (2013) claim that the performance implications of this enrichment should be further evaluated, as data should be moved from the raw source to the real-time cache according to the established time thresholds. Also, we can store multiple data sources, make them available for querying, manipulate them, use them to serve business, and then archive them;

- Adjust the data quality needs according to the latency requirements, i.e., complex data quality jobs take more time to complete than simpler ones focusing on individual values. However, Kimball and Ross (2013) also suggest that value should be added to data as soon as possible, for example using data integration tasks and including results from data mining. There must be a balance between latency and business value;

- Big Data streaming analytics can be relevant for certain data flows, analyzing data and taking actions as it flows through continuous data streams (Kambatla, Kollias, Kumar, & Grama, 2014). In-database analytics can also be a relevant capability to exploit, as Kimball and Ross (2013) highlight.

Begoli and Horey (2012) complement these perspectives, stating that several analytical mechanisms should be included in Big Data solutions, ranging from statistical analysis to data mining and visualization. Moreover, processed data and insights can be made available using open and recognized standards, interfaces, and Web services. Regarding Big Data analytics, there is a vast set of available techniques that can be used to extract value from data. Data mining techniques, such as clustering, association rules, classification, and regression (Han, Pei, & Kamber, 2012), are still present in Big Data environments (Manyika et al., 2011), now with the challenge of distributing them to perform at scale (C. L. P. Chen & Zhang, 2014; Fan & Bifet, 2013). Achieving scalability in these techniques is what makes Big Data analytics different from traditional data analytics. The range of analytical mechanisms and the ambiguous terms to define them may lead to a completely new buzzword: data science. Techniques such as sentiment analysis, time series analysis/ forecasting, spatial analysis, optimization, visualization, or unstructured analytics (e.g., text, audio, and video) (Gandomi & Haider, 2015), can all be part of a data scientist's knowledge base (C. Costa & Santos, 2017b).

2.4.1.2. Architectural and Infrastructural Requirements

The different steps required to process Big Data, presented above, must be performed in Big Data environments, following the requirements identified by Krishnan (2013):

* Absence of fixed data models to adequately accommodate the complexity and size of data, regardless of its characteristics;

* Scalable and high-performance systems to collect and process data either in real-time or in batches;

* The architecture should support data partitioning due to the volume of data;

* Data transformations should use scalable, efficient, and fault-tolerant mechanisms. The results should be stored in adequate systems, such as distributed file systems or non-relational database systems. Data reads should be efficient;

* Data should be replicated and shared across multiple nodes, to support fault-tolerance, multistep processing, and multipartitioning.

Kimball and Ross (2013) corroborate most of the requirements from Krishnan (2013), and also include the following capabilities expected from Big Data environments: possibility to implement User-Defined Functions (UDFs) in several programming languages and to execute them over large datasets within minutes; load and integrate data at high rates; execute queries on streaming data; schedule tasks on large clusters; and support mixed workloads, including *ad hoc* queries or strategic analysis from multiple users, while loading data in batches or in a streaming fashion.

Big Data solutions should be supported by an adequate infrastructure. Regarding this requirement, organizations can currently rely on cloud computing, either by using private, public, or hybrid clouds (Tien, 2013) to provide the underlying resources for massive computations (Hashem et al., 2015). Cloud models, such as Infrastructure-as-a-Service (IaaS), become relevant to accomplish several requirements in Big Data infrastructures, including scalability, commodity hardware, elasticity, fault-tolerance, self-manageability, high throughput, fast I/O, and a high degree of parallelism (Cuzzocrea et al., 2011; Krishnan, 2013). Commodity hardware also plays a relevant role in Big Data infrastructures, namely due to the lower costs of building shared-nothing architectures (Figure 2-5). Google's own papers about the Google File System (GFS) (Ghemawat, Gobioff, & Leung, 2003), MapReduce (Dean & Ghemawat, 2008), and Bigtable (F. Chang et al., 2008) served as inspiration for most of these requirements and for several Big Data technologies that will be presented later.

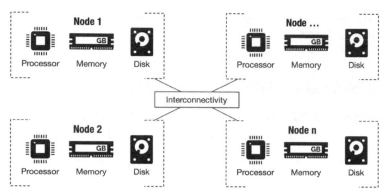

Figure 2-5. A shared-nothing architecture. Adapted from Krishnan (2013).

Kimball and Ross (2013) argue that traditional RDBMSs are not suitable for a wide range of Big Data use cases due to the requirements identified above (e.g., search ranking, sensors, social customer relationship management, document similarity testing, and loan risk analysis). Krishnan (2013) also claims that DWs based on traditional RDBMSs have several design limitations that imply architectural and infrastructural changes to process Big Data, since they cannot be distributed as efficiently as non-relational systems due to Atomicity, Consistency, Isolation, and Durability (ACID) compliance rules, and due to the fact that data partitioning in these systems often does not translate into more scalability or workload reduction. Furthermore, the author mentions the fact that, in many of these systems, the Central Processing Unit (CPU) and memory are often underused, and the way queries are designed typically increases the workload, such as executing queries with a star schema pattern on a 3NF database model, generating significant volume of I/O and inadequate network throughput. Kimball and Ross (2013) present the capabilities that existing RDBMSs vendors are including to extend their solutions for Big Data environments. The authors compare these extended versions with the most commonly recognized open source implementation of MapReduce, namely Apache Hadoop. This comparison is presented in Table 2.2.

Table 2.2. Comparison between an extended RDBMS and Hadoop MapReduce. Adapted from Kimball & Ross (2013).

Characteristic	Extended RDBMS	Hadoop MapReduce
Proprietary	Mostly proprietary	Open source
Cost	Expensive	Less expensive
Variety	Data must be structured	Does not require structure
Type of operations	Adequate for fast indexed lookups	Adequate for massive scans
Relational semantics	Deep support	Indirect support (e.g., Hive)
Complex data structures	Indirect support	Deep support
Transaction processing	Deep support	Little or no support

2.4.2. The Lambda Architecture

The main idea behind the Lambda Architecture (Marz & Warren, 2015) is to think of a Big Data system as a series of layers that satisfy particular needs. As Figure 2-6 shows, the architecture is divided into three main components: batch, serving, and speed layers. In the batch layer, a master dataset stores all the data. Since it is sometimes inefficient to load a dataset that may contain Petabytes of data every time a query is executed, the architecture stores batch views in the serving layer, which are pre-computations of the master dataset. Instead of scanning the entire master dataset, the results are returned from batch views with indexing support, thus obtaining random reads is also possible. Therefore, the batch layer is not only responsible for storing an immutable and constantly growing master dataset, but also for computing functions on it. As Marz and Warren (2015) highlight, creating the batch views is a high latency operation and should be performed in scalable systems. Then, the serving layer stores these batch views in a distributed database supporting batch updates and random reads.

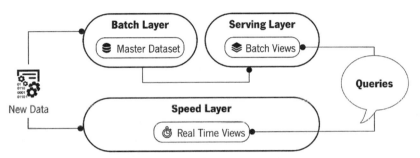

Figure 2-6. The Lambda Architecture. Adapted from Marz & Warren (2015) and C. Costa & Santos (2017a).

However, with only these two layers, batch views would be quickly outdated, as new data takes time to propagate from the batch layer into the serving layer. This does not meet the requirements of low latency (real-time) environments. Consequently, the authors propose the speed layer, which aims to compute functions on data in real-time. Rather than processing all the data at once, like the batch layer, the speed layer only processes recent data. To achieve the smallest possible latency, it does not even look at all the new data at once. Instead, it updates real-time views as new data becomes available, which is described as incremental computation. In order to retrieve current results, queries are answered by looking at the batch and real-time views, merging both results. Consequently, low latency updates are taken into consideration, and as batch views are updated, real-time views can be discarded, since the authors claim that the speed layer is far more complex than the other two. Marz and Warren (2015) describe how to develop Big Data systems according to the principles of the Lambda Architecture, highlighting several technological aspects, as well as other guidelines:

- Store the rawest data to answer as many questions as possible, obtaining different summarizations and insights. Since Big Data technologies are scalable by nature, they can handle this requirement;

- Store untransformed data, since data integration and quality algorithms can be improved in the future;

- Make the master dataset immutable, i.e., only adding more data, without update or delete operations. By doing this, Marz and Warren (2015) claim that human fault-tolerance and simplicity are ensured;

- Within the master dataset, the raw data must be stored as atomic, timestamped, and uniquely identifiable units called facts. The authors describe how to strengthen the fact-based model with information about the types of facts and relationships between them using a graph schema. Moreover, Marz and Warren (2015) also give guidelines about a possible folder and file structure for the master dataset, which would be typically stored in a distributed file system.

2.4.3. Towards Standardization: the NIST Reference Architecture

The NIST Big Data Public Working Group (NBD-PWG), namely the Reference Architecture Subgroup, has been working on an open reference architecture for Big Data (NBD-PWG, 2015), in order to create a tool to facilitate the discussion of requirements, design structures, and operations for Big Data environments. According to the authors, the NBDRA is not a system architecture, but rather a common reference, which is not coupled with specific vendors, services, implementations, or any specific solutions. The NBDRA is presented in Figure 2-7, and the proposed taxonomy for its components is as follows:

- **System orchestrator** – provides requirements regarding policy, governance, architectural design, resources, business requirements, monitoring, and auditing activities. The system orchestrator may include actors such as business leadership, consultants, data scientists, and architects concerned with tasks involving information, software, security, privacy, and network;

- **Data provider** – makes data available through different interfaces, including several data sources (e.g., raw data or previously transformed data). The data provider can be internal or external to the organization;

- **Big Data application provider** – executes the manipulations in the data life cycle to meet the requirements established by the system orchestrator. In this component, several capabilities are combined to create specific data solutions. While the general activities may remain similar to traditional data processing contexts, Big Data methods and techniques are considerably different due to scalability issues;

- **Big Data framework provider** – is composed of general resources or services to be used by the Big Data application provider. According to NBD-PWG (2015), this is the most changed role due to Big Data, taking into consideration the relevance of the infrastructure, data platforms, and processing frameworks. Different technologies can be used and hybrid approaches can emerge, providing flexibility and meeting the requirements of the Big Data application provider;

- **Data consumer** – benefits from the value of the Big Data system. The same type of interfaces used by the data provider can also be exposed to the data consumer, after value has been added to the original data sources.

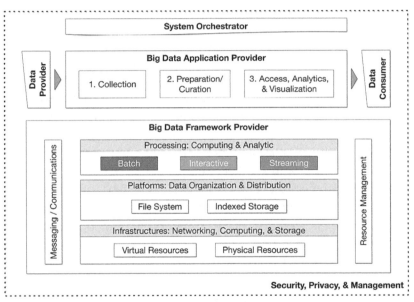

Figure 2-7. The NIST Big Data Reference Architecture. Adapted from NBD-PWG (2015) and C. Costa & Santos (2017a).

The NBDRA has two fabrics encapsulating the aforementioned components: a security and privacy fabric, which affects all the components of the NBDRA and interacts with the system orchestrator (policy, requirements, and auditing), the Big Data application provider, and the Big Data framework provider (development, deployment, and operation); and a management fabric responsible for tasks such as provisioning, software management, or performance monitoring, which involves considerations at scale about the system, data, security, and privacy, while maintaining a high level of data quality and accessibility.

The NBDRA contains five components connected by interoperable interfaces (services) and enveloped by the two fabrics mentioned above. It supports a variety of business environments and facilitates the understanding of how Big Data solutions complement existing approaches and differ from them. To develop this proposal, the authors analyzed a wide range of existing Big Data architectures from the industry, academy, and government (NBD-PWG, 2015).

2.5. BIG DATA TECHNOLOGIES

This subsection highlights several technologies related to Big Data, including Apache Hadoop and related projects, several distributed SQL engines, and other tools for Big Data analytics.

2.5.1. Hadoop and Related Projects

As already mentioned, Hadoop is an open source Apache project based on GFS and MapReduce (Bakshi, 2012). Hadoop has two main components: the Hadoop Distributed File System (HDFS) and a distributed processing framework named Hadoop MapReduce. Hadoop can store and process vast amounts of data by distributing storage and processing across a scalable cluster of multiple nodes built with commodity hardware. In HDFS, files are divided into blocks distributed and replicated across nodes. HDFS ensures that many requirements identified above are fulfilled, such as fault-tolerance and availability, for example. Hadoop MapReduce is a programming model and an execution engine for processing large datasets stored in HDFS, based on the divide and conquer method, dividing a complex problem into many simpler problems, and then combining each simpler solution into an overall solution to the main problem. These are called the Map and Reduce steps (C. L. P. Chen & Zhang, 2014). Regarding HDFS, there are two types of nodes in the cluster: a NameNode, which is responsible for storing metadata about blocks and nodes; and a DataNode, which stores data blocks (Bakshi, 2012). Regarding Hadoop MapReduce, there are also two types of nodes, namely a JobTracker that schedules jobs and distributes tasks across slaves called TaskTrackers (C. L. P. Chen & Zhang, 2014).

Over the years, Hadoop has evolved considerably; changes include the transition from MapReduce to Yet Another Resource Negotiator (YARN) (Hashem et al., 2015). YARN rethinks the JobTracker and TaskTracker components, replacing them with a ResourceManager, a NodeManager, and an ApplicationMaster, to solve some problems in Hadoop MapReduce, such as scalability on large clusters or support for alternative programming paradigms (Krishnan, 2013). Apart from that, Hadoop has several related projects, as Figure 2-8 demonstrates, also highlighting their main features (Apache Hadoop, 2018).

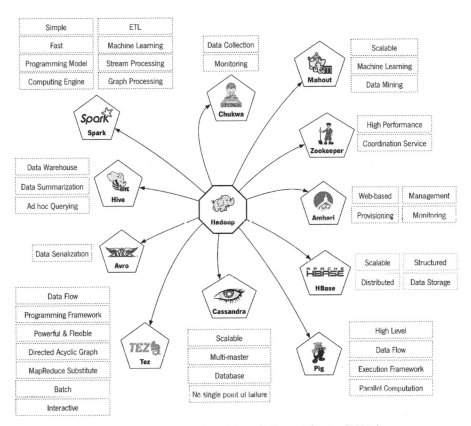

Figure 2-8. The Apache Hadoop ecosystem. Adapted from C. Costa & Santos (2017a).

Other related projects not present in Figure 2-8 may include: Flume, a service to collect, aggregate, and move large amounts of log data; Oozie, a workflow and coordination system for jobs in Hadoop; HCatalog, a metadata layer for data stored in Hadoop, built on top of the Hive Metastore; Sqoop, a connector to integrate data from other existing platforms, such as the DW, metadata engines, enterprise systems, and transactional systems (Krishnan, 2013). There are also more projects that can interact with Hadoop's interfaces or be co-located with it, such as projects for real-time stream processing or interactive *ad hoc* analysis. Since real-time data processing is becoming increasingly relevant to organizations (Chandarana & Vijayalakshmi, 2014), Storm is a real-time computation system to process streams with high throughput and low latency. Kafka, on the other hand, is a messaging/queuing system to produce and consume messages between processes, in an asynchronous and fault-tolerant manner (Marz & Warren, 2015).

Still related to Hadoop, there are several security projects. Hortonworks (2016) identifies five pillars for security in Hadoop: administration, authentication, authorization, auditing, and data protection. Kerberos, Apache Knox, and Apache Ranger are highlighted as projects related to these five pillars, in order to ensure a secure Hadoop environment. Kerberos can be used to authenticate users and resources within Hadoop clusters. Apache Knox complements Kerberos, by blocking services at the perimeter of the cluster and hiding the cluster's access points from end users, thus adding another layer of protection for perimeter security. Finally, Ranger provides a centralized platform for policy administration, authorization, auditing, and data protection (e.g., encrypted files in HDFS).

2.5.2. Landscape of Distributed SQL Engines

Interactive and low latency *ad hoc* analysis over large datasets is a relevant scenario in organizations. Occasionally, users do not know the queries in advance and need to execute *ad hoc* queries within milliseconds or seconds, even at scale. There is a trend named SQL-on-Hadoop (Floratou, Minhas, & Özcan, 2014) that is related to the implementation of distributed SQL engines for interactive *ad hoc* analysis of large datasets stored not only in Hadoop, but also in distributed databases (e.g., NoSQL databases, described in more detail in section 3.2). Many SQL-on-Hadoop systems are available under open source licenses, including: Hive on Tez (Huai et al., 2014); Presto (2018); Impala (Kornacker et al., 2015); Drill (Hausenblas & Nadeau, 2013); HAWQ (L. Chang et al., 2014); and Spark SQL (Armbrust et al., 2015), among many others. These systems are able to combine data from multiple sources like HDFS files, Hive tables (section 4.1), NoSQL databases (section 3.2), SQL databases, and Kafka, for example, which means that in a single query they can combine not only data from different systems, but also batch and streaming data. Consequently, SQL-on-Hadoop systems play a relevant role in BDWing systems, as shown later in this book. A brief description of the main characteristics of these tools is provided in this subsection, whereas a more detailed description of these tools can be found in Rodrigues, Santos, and Bernardino (2018).

Drill (2018) is a system that supports data-intensive distributed applications for interactive analysis of large-scale datasets. It provides low latency and fast processing for interactive queries. Drill supports several file systems and databases, like HDFS, Hive, or HBase, and uses SQL to query data. It can scale-up to a significantly large cluster with thousands of nodes and offers connectivity and compatibility with BI tools like Tableau, MicroStrategy, QlikView, and Tibco. Among its various components, the core is named Drillbit. This service is responsible for accepting requests from the clients, processing the queries, and returning the results to the clients (Rodrigues et al., 2018).

HAWQ (2018), Hadoop With Query, is a Hadoop native SQL engine that combines Massively Parallel Processing (MPP) with the scalability of Hadoop, delivering adequate performance when querying large datasets. For performing complex queries, HAWQ splits the queries into small tasks and distributes them to the MPP execution units (Szegedi, 2014). Applications and external tools can interact with HAWQ using standard protocols such as Open Database Connectivity (ODBC) or Java Database Connectivity (JDBC). In addition, HAWQ can be used with MADlib (Pivotal, 2017), an open source library for scalable in-database analytics, making machine learning capabilities available through a SQL interface.

Hive (2018a) supports the processing and analysis of data stored in HDFS and it is considered the DW storage system for Big Data. Hive was the first processing tool supporting SQL-on-Hadoop and has been used by many organizations, such as Amazon, to store and process large amounts of data. Being frequently considered as a high-latency system, oriented for batch workloads instead of interactive querying, Hive has improved its performance adopting new storage formats like the Optimized Row Columnar (ORC) file format (described in more detail in subsection 4.1.1) and the Tez execution engine, optimizing the execution of Hive jobs (Gates, 2014). Hive gives structure to data stored in HDFS and performs *ad hoc* querying and analysis using the Hive Query Language (HiveQL), a SQL-like query language.

Impala (2018) is an MPP SQL query engine that provides adequate performance, real-time processing, low latency, and high concurrency processing. Impala combines SQL-based queries with multi-user performance in Hadoop environments, which can include hundreds of machines. Data can be stored in HDFS or HBase, using the same metadata, SQL syntax, ODBC driver, and user interface as Hive, providing a unified platform for batch-oriented and real-time query processing.

Presto (2018) is a distributed SQL query engine for running interactive analytical queries. It was developed by Facebook and it is optimized for low latency and interactive queries. Presto supports several SQL features including joins, aggregations, and subqueries, as well as other features that include the manipulation of JSON documents, URL functions, strings, and regular expressions. A query in Presto can integrate data from multiple sources, allowing data analytics across several data sources, meaning it can query data available in HDFS, as well as in other data sources, including relational or non-relational databases.

Spark (2018) is a highly distributed processing framework that provides efficient analytics on heterogeneous data. Spark provides full access and compatibility with existing Hive data and uses the concept of Resilient Distributed Dataset (RDD), which describes an immutable collection of objects that are partitioned and distributed across multiple nodes in a cluster, allowing for distributed processing. Spark includes Spark SQL, a module for processing structured data with DataFrames/Datasets (equivalent to tables in relational databases), organizing data in named columns.

Given the specificities of each tool, Table 2.3 presents a brief comparison of these SQL-on-Hadoop technologies (Rodrigues et al., 2018), looking into the following set of characteristics, considered as main requirements in a Big Data processing architecture (Grover et al., 2015; Jethro, 2018; Landset, Khoshgoftaar, Richter, & Hasanin, 2015; MapR, 2018), such as SQL support, latency, scalability, processing performance, supported use cases, points of failure, fault-tolerance, and support for machine learning or data visualization tools.

Table 2.3. SQL-on-Hadoop systems. Adapted from Grover et al. (2015); Jethro (2018); Landset et al. (2015); MapR (2018); and Rodrigues et al. (2018).

Feature	Drill	HAWQ	Hive	Impala	Presto	Spark SQL
SQL support	ANSI SQL	ANSI SQL	HiveQL	HiveQL	ANSI SQL	ANSI SQL (limited) & HiveQL
Latency	Low	Low	Medium	Low	Low	Low
Scalability	High	High	High	Very high	Very high	High
Processing speed	Fast	Very fast	Fast	Very fast	Very fast	Fast
Use cases	Real-time interactive queries and analysis/BI	Batch processing and interactive queries	Batch processing	Real-time interactive queries and analysis/BI	Real-time interactive queries	Batch and stream processing, interactive queries and machine learning
Single point of failure in query execution	No	Yes	Yes, if failure at the master node	Yes, if any host quits query execution	Yes, if any host quits query execution	No
Fault-tolerance	Yes	Yes	Yes	No	No	Yes
Machine learning	No	Yes	No	Yes	No	Yes
Data visualization	No native support, compatibility with BI tools	No native support, compatibility with BI tools	No native support, compatibility with BI tools	No native support, compatibility with BI tools	No native support, compatibility with BI tools	No native support, compatibility with BI tools

Moreover, besides SQL-on-Hadoop systems, there are other similar technologies targeting interactive *ad hoc* querying, such as Druid (see section 4.3), a columnar store that provides real-time aggregation and indexing at data ingestion time (Yang et al., 2014).

2.5.3. Other Technologies for Big Data Analytics

By describing Hadoop and its related projects, several technologies for Big Data analytics were already inherently identified: streaming analytics (e.g., Spark Streaming and Storm); data mining and machine learning (e.g., Spark and Mahout); DWing (e.g., Hive); interactive *ad hoc* analysis (e.g., Hive on Tez, Impala, Presto, Drill, and Spark SQL) (Santos et al., 2017; Soliman, 2017); data flow (e.g., Pig). However, no data visualization tools have been introduced yet.

Regarding Big Data visualization, several mashup tools can be highlighted, such as Datameer, FICO Big Data Analyzer (former Karmasphere), Tableau, and TIBCO Spotfire (Krishnan, 2013). These mashup tools can bring together data from multiple sources into a single picture. As Krishnan (2013) highlights, there is also the possibility of visualizing Big Data with statistical tools, like R or SAS, for example, taking advantage of their capabilities. Other tools are also briefly mentioned in the literature, such as Jaspersoft Business Intelligence (BI) Suite and Pentaho Business Analytics (C. L. P. Chen & Zhang, 2014). Certainly, many other visualization tools exist and may be adequate for Big Data visualization, such as Excel and Power BI, JavaScript libraries, or Python's plot capabilities.

Besides data visualization, there are other tools to extract, load, transform, and integrate data before carrying out analytical tasks. Talend Open Studio for Big Data is an example of such tool (C. L. P. Chen & Zhang, 2014). Moreover, apart from the aforementioned tools related to Hadoop for data mining and machine learning, other alternatives identified in the literature include: MADLib and EMC Greenplum (Begoli & Horey, 2012); R, MOA, WEKA, and Vowpal Wabbit (Fan & Bifet, 2013); data mining tools from SAS or IBM (Krishnan, 2013); Rapidminer; and KNIME (M. Chen et al., 2014). Some of these tools, like R and WEKA, are not scalable by default, and they are also used in traditional data mining and machine learning environments, where processing large training sets is not a significant concern. Over time, these tools have been extended with several connectors for scalable Big Data stores and packages for distributed processing (e.g., SparkR, RHadoop, RHive, distributedWekaBase, distributedWekaHadoop, and distributedWekaSpark), but, by default, without these extensions, they are better suited for small to moderate datasets. This does not mean that they are not useful in Big Data mining, quite the opposite, but the volume of data that serves as training and testing sets should be considered (preprocessing large datasets can be useful in these cases). The same principle applies to other non-distributed algorithms implemented in any other languages, like Python or Java, for example. It should be remembered that one of the main challenges regarding the Big Data life cycle is scaling the algorithms to extract value from data (Hashem et al., 2015).

3 OLTP-ORIENTED DATABASES FOR BIG DATA ENVIRONMENTS

INTRODUCTION

Over the last years, the NoSQL movement has revolutionized the way we think about databases that support large-scale enterprise applications. NoSQL databases can be classified according to different types and can also have various data models, being frequently recognized as schema-less storage systems capable of serving low latency applications that handle huge amounts of data. This chapter presents an overview of the NoSQL movement and the various types of NoSQL databases, including practical examples for Redis (key-value store), HBase (wide column store), MongoDB (document store), and Neo4j (graph database). Moreover, the chapter concludes with some insights regarding NewSQL databases, a relatively recent movement focused on bridging the gap between SQL and NoSQL databases by trying to combine the advantages of both.

3.1. NOSQL AND NEWSQL: AN OVERVIEW

Database technology has evolved significantly to handle datasets at different scales and support applications that may have high needs for random access to data (M. Chen et al., 2014; Hashem et al., 2015). NoSQL databases became popular mainly due to the lack of scalability in RDBMSs, since this new type of databases provides mechanisms to store and retrieve large volumes of distributed data (Hashem et al., 2015). The relevant factors that motivated the appearance of NoSQL databases were the strictness of the existing relational model and, consequently, its inadequacy for storing Big Data. NoSQL databases are regarded as distributed, scalable, elastic, and fault-tolerant storage systems. They satisfy an application's need for high availability, even when nodes fail, appropriately replicating data across multiple machines (Kambatla et al., 2014). Relational databases will certainly evolve and some organizations, such as Facebook, are using mixed database architectures (M. Chen et al., 2014). Combining the benefits of both storage systems is a current research trend, as was previously mentioned (Cuzzocrea et al., 2011). A recent term is emerging, "NewSQL", which combines the relational data model with certain benefits of NoSQL systems, such as scalability, for example (Grolinger, Higashino, Tiwari, & Capretz, 2013). According to Cattell (2011), NoSQL and NewSQL databases are mainly designed to scale OLTP-style workloads over several nodes, hence fulfilling the requirements of environments with millions of simple operations (e.g., key lookups and reads/writes of one record or a small number of records).

This phenomenon changed the way databases are now designed. While a RDBMS complies to ACID properties (Krishnan, 2013), a NoSQL database, being a distributed system, typically follows the principle of the Consistency, Availability, Partition tolerance (CAP) Theorem: *"any networked shared-data system can have only two of three desirable properties"* (Brewer, 2012, p. 23). In the case of the NoSQL databases, these properties include: consistency, equivalent to a single up-to-date copy of the data; high availability of that data; and tolerance to network partitions. As Brewer (2012) claims, CAP served its purpose of leveraging the design of a wider range of systems and trade-offs, and the NoSQL movement is a clear example of that. The fact that two of the three properties should be chosen was always misleading, states the author, as it tends to simplify the "tensions among properties". These properties are more continuous than binary and, therefore, they can have many intermediate levels. CAP only proscribes perfect availability and perfect consistency in the presence of network partitions (Figure 3-1).

Consequently, the CAP Theorem serves to consider combinations of consistency and availability that fit a given scenario. The choices between consistency and availability can vary within a certain system, for example, according to specific data or users. Brewer (2012) clarifies this misunderstanding and discusses the relationship between ACID and CAP, stating that choosing availability only affects some of the ACID's guarantees. These design considerations are intrinsic to NoSQL databases and each may be designed differently regarding these choices.

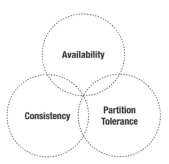

Figure 3-1. The CAP Theorem. Adapted from Hewitt (2010).

There are many NoSQL databases, so enumerating and evaluating all of them is virtually impossible. Tudorica and Bucur (2011) state that over 120 NoSQL databases were known in 2011. In addition, other sources point to the existence of more than 225 NoSQL databases (NoSQL, 2018). This number increases so fast that it is almost impossible to follow the evolution of databases in terms of their characteristics, advantages, and disadvantages.

Regarding their types, NoSQL databases are usually divided into four data models, which are described as follows along with examples (more detailed explanations are provided in the next subsections of this chapter):

- **Key-value model** – data is typically stored in key-value pairs. The key uniquely identifies a value of an arbitrary type. These data models are known for being schema-free, but may lack the capability to adequately represent relationships or other structures, since queries and indexing depend on the key (Grolinger et al., 2013). Every key is unique and queries are tightly coupled with the keys (M. Chen et al., 2014). Examples: Redis, Memcached, BerkeleyDB, Voldemort, Riak, and Dynamo;

- **Column-oriented model** – a column-oriented data model can be seen as an extension of the key-value model that adds columns and column families, thus providing more powerful indexing and querying (Krishnan, 2013). This design was largely inspired by Bigtable (M. Chen et al., 2014; Grolinger et al., 2013), although not all column-oriented databases are completely inspired by it (e.g., Cassandra adopts design aspects from both Dynamo and Bigtable). Examples: Bigtable, HBase, Cassandra, and Hypertable;

- **Document-oriented model** – suitable for representing data in a document format, typically using JSON. It can contain complex structures, such as nested objects, and also typically includes secondary indexes, thus providing more query flexibility than the key-value data model (Grolinger et al., 2013). Examples: MongoDB, CouchDB, and Couchbase;

- **Graph model** – based on graph theory, in which objects can be represented as nodes, and relationships between them can be represented as edges (Krishnan, 2013). Graphs are specialized in handling interconnected data with several relationships (Grolinger et al., 2013). Examples: Neo4j, InfiniteGraph, GraphDB, AllegroGraph, and HyperGraphDB.

These different NoSQL database types vary in terms of the data models they support, as was previously mentioned, but also in the data volume each one is most suited to handle. When the complexity of the data models increases, the data volume each database can handle decreases. As shown in Figure 3-2, key-value databases are more appropriate for large volumes of non-complex data, which are stored as key-value pairs. Each value can range from a single attribute to a list of attributes without any predefined relationship between them. As data models begin to include relationships between attributes or some type of data organization (like columns in tables or nested structures in documents), the complexity of the data models increases, hence decreasing the capability to handle more data volume. NoSQL graph databases are considered the most appropriate for storing and processing relationship-heavy data.

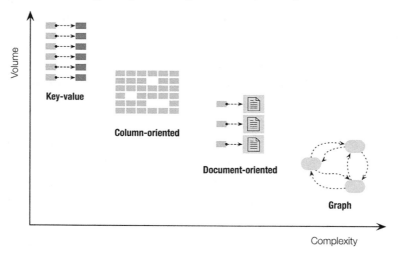

Figure 3-2. Types of NoSQL databases. Adapted from Hewitt (2010).

NewSQL databases are based on the relational model (Grolinger et al., 2013), offering either a pure relational view of the data (e.g., VoltDB, Clustrix, NuoDB, MySQL Cluster, ScaleBase, ScaleDB) or something similar (e.g., Google Spanner). According to Grolinger et al. (2013), sometimes interactions with these databases rely on tables and relations, but they might use different internal representations. Such is the case of NuoDB. Different NewSQL databases support different SQL compatibility, such as unsupported clauses or other incompatibilities with the standard. Similar to NoSQL,

NewSQL databases can scale by adding nodes to a cluster, using a horizontal scaling approach (Figure 3-3).

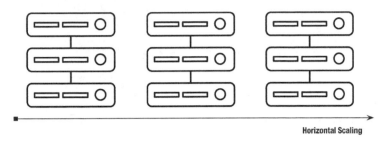

Horizontal Scaling

Figure 3-3. Horizontal scaling of nodes in a cluster.

3.2. NOSQL DATABASES

The following subsections describe in more detail each one of the different types of NoSQL databases, providing some guidelines about their data models and some examples of their use.

3.2.1. Key-value Databases

This subsection provides an overview of the key-value stores and presents the Redis database in more detail, with some examples of its data model and functionalities.

3.2.1.1. Overview

Key-value stores, or key-value databases, are the most flexible storage mechanism available in NoSQL databases, being an alternative to traditional databases with pre-defined data schemas. This storage paradigm relies on the concepts of key and value. The key can be mapped to a value or to a complex version of a value, which may include a list of values for several attributes (Sadalage & Fowler, 2012).

The keys can be stored in a hash table, or similar structure, and the values are identified and accessed using the keys (Figure 3-4). Values can include a wide variety of data types or data structures, expressed using numbers, strings, images, videos, maps, or JSON, XML, and Hypertext Markup Language (HTML) objects. Since there is no pre-defined data model or data structure, the application layer provides the semantics for what is stored in the values. This flexibility implies an overhead effort in developing the application layer for processing and analyzing the stored data.

© FCA

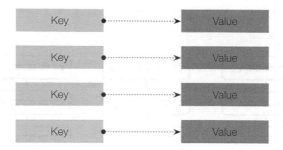

Figure 3-4. Key-value stores.

The data model of a key-value store is based on these two components, the key and the value, where the value is only available for a lookup after knowing the corresponding key. Constrained by the specific selected database, some size limitations may exist for the value component. Apart from those limitations, there is complete freedom to store data in the value. This type of repository is optimized for querying using the keys. As shown in Figure 3-5, the value component may vary from record to record, and this also applies to the number of elements or attributes.

Figure 3-5. Example of a key-value store.

This type of database enables high performance and availability, with efficient query processing and easy distribution of data across a cluster; however, it does not support processing of complex query filters, join operations, or foreign key constraints. Consistency can only be verified in operations on single keys.

3.2.1.2. Redis

Redis (2018b) is a non-relational database that stores a mapping of keys pointing to values. Its key-value approach supports data structures with different values such as strings, hashes, lists, sets, stored sets, bitmaps, and hyperLogLogs. The five fundamental data structures of Redis are shown in Figure 3-6. In them, a string key can map a string value or a more complex data structure, such as:

- **Lists** – collections of string elements stored by order of insertion;

- **Sets** – collections of unique and unsorted string elements;

- **Hashes** – maps composed of fields and their corresponding values, in which both fields and values are strings;

- **Sorted sets** – sets where each string element includes a score represented by a floating number value, allowing the retrieval of a range of elements, such as five of the top or bottom elements. This kind of selection is not possible in sets.

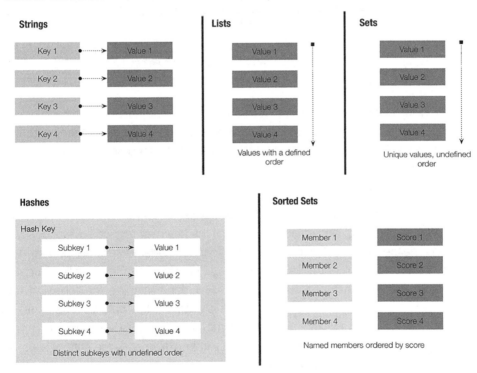

Figure 3-6. Examples of Redis data structures. Adapted from Redis (2018a).

Besides these five fundamental data structures, Redis allows:

- **Bit arrays or bitmaps** – arrays of bits that can be used to handle string values. Individual bits can be set or cleared, can be counted (0's or 1's), can be searched, among other special operations;

- **HyperLogLogs** – a probabilistic data structure that can be used to count the number of distinct values and obtain the cardinality of a set.

Regarding the fundamental data structures, Table 3.1 summarizes the values each structure can handle, as well as the common operations for that structure.

Table 3.1. Data structures in Redis. Adapted from Redis (2018a).

Data Structure	Values	Operations
Strings	Strings, integers, or floating-point values	Operate on strings; increment/decrement integers and floats
Lists	Lists of strings	Push or pop items; read individual or multiple items; find or remove items by value
Sets	Unordered collection of unique strings	Add, fetch, or remove individual items; check membership, intersect, union, or difference; fetch random items
Hashes	Unordered hash table of keys to values	Add, fetch, or remove individual items; fetch the whole hash
Sorted sets	Ordered mapping of string members with floating-point scores, ordered by score	Add, fetch, or remove individual values; fetch items based on score ranges or member value

Redis is an in-memory storage solution with the capability of persistent storage on disk, which supports replication to scale read performance, and client-side sharding to scale write performance. This sharding option partitions the data based on some component of the key (such as an ID) and/or based on the hash of the keys, storing and fetching data from multiples nodes in a cluster (Redis, 2018a).

The keys in Redis are binary safe, meaning that any binary sequence can be used as a key. Keys must not be too long, because the lookup of the key may require several costly key comparisons, neither too short to save storage space. Readable keys are preferred and usually do not add a significant amount of space. It is recommended to follow a pattern when defining keys, for instance, `object-type:id`, for assigning the following keys: `student:0001`, `student:0002`, `student:0003`, etc. When the data volume increases, maintaining unique values for keys can be difficult, it is therefore advisable to define a pattern for generating strings that remain unique within extremely large sets of keys. The maximum key size is 512MB (Redis, 2018b).

To provide an overview of Redis' data structures and the operations that can be performed on them, some illustrative examples are now offered. Starting with strings, three manipulation commands are available, as shown in Table 3.2.

Table 3.2. Redis commands for strings. Adapted from Redis (2018a).

Command	Action
`get`	Fetches the data associated with a key
`set`	Sets the value of a given key
`del`	Deletes the value of a given key

Using `redis-cli`, the Redis Command Line Interface (CLI), the following listing code presents an example where the value `myValue` is assigned to the `v001` key. The value is then fetched and deleted using the key. After the removal, Redis returned the `nil` value when trying to fetch the value again (Redis, 2018a).

```
$ redis-cli
redis> set v001 myValue
OK
redis> get v001
"myValue"
redis> del v001
(integer) 1
redis> get v001
(nil)
```

In lists, Redis supports a linked-list structure in which it is possible to store a sequence of strings as the value of a key. The available operations allow to push items to the front or to the back of the list, to pop items from the front or from the back of the list, to fetch an item at a given position or to fetch a range of items (Table 3.3).

Table 3.3. Redis commands for lists. Adapted from Redis (2018a).

Command	Action
`lindex`	Fetches an item available at a given position in the list
`lpop/rpop`	Pops the value from the left/right end of the list and returns it
`lrange`	Fetches a range of values from the list
`lpush/rpush`	Pushes the value into the left/right end of the list

As an example, the following listing code shows how the `list-001` list is filled with three items (`item01`, `item02`, `item03`) that are inserted at the right with a `rpush`. As the items are incrementally inserted, Redis returns the current length of the list. Afterwards, the entire list is fetched with `lrange`, using `0` for the start index and `-1` for the last index. The `lindex` command is used to fetch an item from the list, namely the one referenced by index `1`. The removal of an item from the list can be done with `lpop`, which in this case is deleting `item01` from the list. Using again the `lrange` command shows the two remaining items in the list. Besides the generic example on the left side of the listing code (Redis, 2018a), a specific one is shown on the right side of the listing code, indexed to a "`countries`" instance.

© FCA

```
redis> rpush list-001 item01          redis> rpush list-countries Portugal
(integer) 1                           (integer) 1
redis> rpush list-001 item02          redis> rpush list-countries Spain
(integer) 2                           (integer) 2
redis> rpush list-001 item03          redis> rpush list-countries France
(integer) 3                           (integer) 3
redis> lrange list-001 0 -1           redis> lrange list-countries 0 -1
1) "item01"                           1) "Portugal"
2) "item02"                           2) "Spain"
3) "item03"                           3) "France"
redis> lindex list-001 1              redis> lindex list-countries 1
"item02"                              "Spain"
redis> lpop list-001                  redis> lpop list-countries
"item01"                              "Portugal"
redis> lrange list-001 0 -1           redis> lrange list-countries 0 -1
1) "item02"                           1) "Spain"
2) "item03"                           2) "France"
```

Sets are very similar to lists, but the sequence of strings is managed by a hash table that ensures all strings are unique. Because sets are unordered, it is not possible to push or pop items from the top or bottom sections, as in lists. However, items can be added or removed, items can be searched to verify if they exist in the set, or the entire set can be fetched (Table 3.4).

Table 3.4. Redis commands for sets. Adapted from Redis (2018a).

Command	Action
sadd	Adds an item to the set
smembers	Returns all the items in the set
sismember	Checks if an item belongs to the set
srem	Removes the item from the set, if it exists

To illustrate how sets work, the following listing code starts by adding three items (item01, item02, item03) to the set-001 set. For each new successfully inserted item, the returned value is 1. If the item already exists in the set, the returned value is 0. Then, smembers is used to fetch all the items within that set. To verify if an item is in a set, sismember returns 1 or 0 for yes and no, respectively. To remove items, srem is used, returning 1 or 0 when the item is successfully removed (for items that do exist in the set), or not, respectively. The following listing code shows the generic example (Redis, 2018a), alongside one indexed to a "countries" instance.

```
redis> sadd set-001 item01 item02
item03
(integer) 3
redis> sadd set-001 item03
(integer) 0
redis> smembers set-001
1) "item01"
2) "item02"
3) "item03"
redis> sismember set-001 item07
(integer) 0
redis> sismember set-001 item02
(integer) 1
redis> srem set-001 item03
1
redis> srem set-001 item03
0
redis> smembers set-001
1) "item01"
2) "item02"
```

```
redis> sadd set-countries Portugal
Spain France
(integer) 3
redis> sadd set-countries France
(integer) 0
redis> smembers set-countries
1) "Portugal"
2) "Spain"
3) "France"
redis> sismember set-countries Greece
(integer) 0
redis> sismember set-countries Spain
(integer) 1
redis> srem set-countries France
1
redis> srem set-countries France
0
redis> smembers set-countries
1) "Portugal"
2) "Spain"
```

In Redis, hashes are used for mapping keys to values, which can be string values or numeric values. In this last case, numeric values can be incremented or decremented. An hash can be seen as a set of pairs, where the hash includes subkeys to index its several values (Redis, 2018a).

Regarding commands, hashes allow the storage of the keys, fetching the value of a given hash key, fetching the entire hash and removing a key from the hash in case it exists (Table 3.5).

Table 3.5. Redis commands for hashes. Adapted from Redis (2018a).

Command	Action
hset	Stores a value in a hash key
hget	Fetches the value of a given hash key
hgetall	Fetches the whole hash
hdel	Removes a key from the hash, if the key exists

In the following example, the `h001` hash is created using the `hset` command with three elements. The subkeys of these elements follow a pattern that uses the hash label and a number as a sub-label, only to show the hierarchical dependence between the hash and its members (`h001-001`, `h001-002`, ...). The `hgetall` shows all the subkeys and values included in the hash, whereas the `hdel` is used to delete a specific member in the hash

© FCA

structure and the `hget` is used to fetch a specific value for a subkey in the hash. Along the generic example (Redis, 2018a), a specific one is shown indexed to a "`countries`" instance.

```
redis> hset h001 h001-001 item01
(integer) 1
redis> hset h001 h001-002 item02
(integer) 1
redis> hset h001 h001-003 item03
(integer) 1
redis> hset h001 h001-003 item03
(integer) 0
redis> hgetall h001
1) "h001-001"
2) "item01"
3) "h001-002"
4) "item02"
5) "h001-003"
6) "item03"
redis> hdel h001 h001-002
(integer) 1
redis> hget h001 h001-003
"item03"
redis> hgetall h001
1) "h001-001"
2) "item01"
3) "h001-003"
4) "item03"
```

```
redis> hset h-countries pt Portugal
(integer) 1
redis> hset h-countries sp Spain
(integer) 1
redis> hset h-countries fr France
(integer) 1
redis> hset h-countries fr France
(integer) 0
redis> hgetall h-countries
1) "pt"
2) "Portugal"
3) "sp"
4) "Spain"
5) "fr"
6) "France"
redis> hdel h-countries sp
(integer) 1
redis> hget h-countries fr
"France"
redis> hgetall h-countries
1) "pt"
2) "Portugal"
3) "fr"
4) "France"
```

The last data structure, sorted sets or zsets, is similar to hashes in the sense that the several keys (members of the sorted set) are unique and point to a value (called score) that is a floating-point number. The several members can be accessed using the key, which corresponds to the member itself, or can be accessed by the sorted order or by the values of the scores (Redis, 2018a). Members can be added, removed, or fetched from a sorted set (Table 3.6).

Table 3.6. Redis commands for sorted sets. Adapted from Redis (2018a).

Command	Action
zadd	Adds a member with a given score to a sorted set
zrange	Fetches a member in the sorted set, given its position in the sorted order
zrangebyscore	Fetches members in the sorted set, given a range of scores
zrem	Removes a member from the sorted set, if it exists

In the following example, the `zset-001` sorted set is created using `zadd` with three members, `item01`, `item02`, and `item03`, each one with its respective score, `7`, `12` and `15`. The `zrange` fetches all the members of the sorted set, and the `zrangebyscore` fetches all the members inside the interval of scores. For deleting a specific member, `zrem` is used. Besides the generic example (Redis, 2018a), a specific one is shown indexed to a `"countries"` instance.

```
redis> zadd zset-001 7 item01
(integer) 1
redis> zadd zset-001 12 item02
(integer) 1
redis> zadd zset-001 15 item03
(integer) 1
redis> zadd zset-001 15 item03
(integer) 0
redis> zrange zset-001 0 -1
withscores
1) "item01"
2) "7.0"
3) "item02"
4) "12.0"
5) "item03"
6) "15.0"
redis> zrangebyscore zset-001 0 10
"item01"
redis> zrem zset-001 item03
1
redis> zrange zset-001 0 -1
withscores
1) "item01"
2) "7.0"
3) "item02"
4) "12.0"
```

```
redis> zadd z-countries 17 Portugal
(integer) 1
redis> zadd z-countries 54 Spain
(integer) 1
redis> zadd z-countries 75 France
(integer) 1
redis> zadd z-countries 75 France
(integer) 0
redis> zrange z-countries 0 -1
withscores
1) "Portugal"
2) "17.0"
3) "Spain"
4) "54.0"
5) "France"
6) "75.0"
redis> zrangebyscore z-countries 0 20
"Portugal"
redis> zrem z-countries France
1
redis> zrange z-countries 0 -1
withscores
1) "Portugal"
2) "17.0"
3) "Spain"
4) "54.0"
```

The previous sections introduced the main concepts of key-value stores along with brief examples of the use of Redis, a specific key-value database. For further details, see the Redis detailed documentation available at Redis (2018a; 2018b).

3.2.2. Column-oriented Databases

This subsection provides an overview of the column-oriented stores and describes HBase in more detail, along with some examples of its data model and functionalities.

3.2.2.1. Overview

As the name implies, in a column-oriented store, or column-oriented database, data is stored using a column-oriented model, which, depending on the physical organization of the data, may be very efficient for compressing it or processing it using aggregation operations (George, 2010). One of the possible physical views of the data that ensures these capabilities is based on storing data from the same column together as a continuous disk entry, hence the reason why it is very efficient for compression or even for data processing. Read and write operations are done using columns, not the traditional row-oriented organization strategy found in most relational databases (George, 2010).

This is one of the main physical views of the data found in column-oriented databases. However, since the term column-oriented is widely used, sometimes to describe different physical views of the data, not all "column-oriented" databases fit under this strategy to physically store the data. Some column-oriented databases may physically store the columns as a continuous disk entry, others on a per column family basis (groups of columns), among other strategies. For this reason, very often, the term "column-oriented" is used as an umbrella both for pure columnar physical views of the data and for other strategies found on a set of databases typically named "wide column stores" that, despite being often tagged as "column-oriented", frequently store the data on a row-by-row fashion that allows for different rows to have different columns, on a highly dynamic manner, hence the coined term "schema-less".

Consequently, whenever practitioners see a database labeled as "column-oriented", further clarification is needed regarding the way it physically stores the data, since depending on the strategy, different databases leverage different aggregation, random access, and compression capabilities. For example, databases that store columns as a continuous disk entry provide excellent sequential access capabilities for efficient aggregations, while more row-oriented wide column stores allow for very efficient random writes or random reads.

Looking at the table shown in Figure 3-7, the row-oriented data model groups the data by row, whereas the column-oriented data model groups the data by column, taking advantage of data types for compacting and processing data (George, 2010).

At the conceptual level, one of the typical logical views for the data model of this type of databases relies on a **keyspace** as the outermost data organization level, representing a specific database. Tables inside a **keyspace** include a **key**, which is a unique value that is used for the identification of a specific record. Tables include columns that can be grouped into column families. Although the available elements (**keyspace, columns, column families**, among others) will depend on the chosen database, many column-oriented stores use this fundamental set of elements or some subset of it.

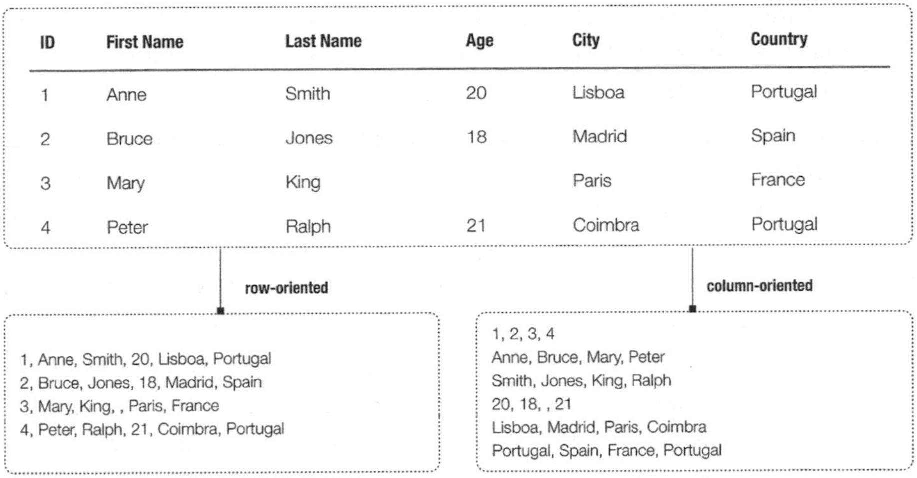

ID	First Name	Last Name	Age	City	Country
1	Anne	Smith	20	Lisboa	Portugal
2	Bruce	Jones	18	Madrid	Spain
3	Mary	King		Paris	France
4	Peter	Ralph	21	Coimbra	Portugal

row-oriented

1, Anne, Smith, 20, Lisboa, Portugal
2, Bruce, Jones, 18, Madrid, Spain
3, Mary, King, , Paris, France
4, Peter, Ralph, 21, Coimbra, Portugal

column-oriented

1, 2, 3, 4
Anne, Bruce, Mary, Peter
Smith, Jones, King, Ralph
20, 18, , 21
Lisboa, Madrid, Paris, Coimbra
Portugal, Spain, France, Portugal

Figure 3-7. Row-oriented vs. column-oriented storage.

In the case of column-oriented databases that allow column families, each column belongs to one column family, which is used for reading and writing, since data from a specific column will be usually accessed together (Sadalage & Fowler, 2012). Different rows can have data from different columns, since most databases allow adding new columns over time without the need to fill in default values for the existing rows in the new cells. Data schemas can change over time, accommodating new data requirements.

The following subsection presents in more detail the concept of column-oriented database and the available elements, using as an example the HBase NoSQL database.

3.2.2.2. HBase

A well-known column-oriented database is HBase (2018a), a distributed database built on Hadoop and its distributed file system. In this context, column-oriented means that HBase makes use of an on-disk column family-oriented storage format. With it, HBase provides key-based access to a specific cell of data or to a sequential range of cells (George, 2010). Moreover, HBase can be classified as a column family-oriented storage, as it groups columns into column families (George, 2010). Inside each column family, data is stored in a row-oriented format, as the examples of this subsection will show. Next chapter (subsection 4.1.1) will further detail these differences when presenting the several data storage formats.

HBase allows real-time read/write random access to very large datasets and is designed to scale linearly just by adding nodes to a cluster (horizontal scalability). HBase provides fault-tolerant storage of large quantities of data in a cluster made of commodity hardware, with high availability and high performance (White, 2015).

Since it is non-relational, HBase does not support SQL, although nowadays the Apache Phoenix project (Phoenix, 2018) provides a layer that implements a transactional SQL database on top of HBase (2018b).

Regarding the logical data model, data is stored in **tables** integrating **rows** and **columns**. Columns for a specific instance, or row, are addressed uniquely by a **row key**. All the **cells** in a table are automatically versioned by HBase, adding an auto-assigned **timestamp** when the cell is inserted (the timestamp can also be explicitly set by the user). The cell content is an array of bytes. Tables are grouped into **namespaces** that improve the organization of the data supporting a specific application, for instance.

All row keys must be unique because these keys are used to sort the data in a lexicographical manner. In this sorting process, each key is compared on a binary level, one byte at a time, from left to right. The resulting sorted set provides a kind of primary key index that enhances data management tasks, such as updating or searching for data.

Columns in tables are grouped into **column families**, which adds a semantic layer useful not only to model the data but also to physically organize it, as all columns in a column family are stored together in the same low-level storage file: the HFile (HBase, 2018b). Column families are defined when the table is created, and it is advisable not to change them too often nor defining too many of them in a table.

To exemplify, Figure 3-8 shows a table of students along with two column families: name and address; each one holding two columns which define the students' first and last names and their city and country, respectively. In this table, it is possible to see the row key and the timestamp that is automatically assigned by the system. Afterwards in this chapter, more details about the creation of the tables' key will be given, taking into consideration the relevance of the key to maintain an adequate performance for data manipulation tasks. For now, the examples maintain the keys as abstract and simple as possible.

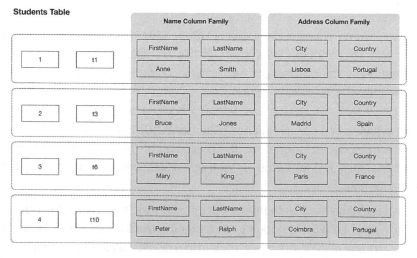

Figure 3-8. Example of a HBase table.

HBase tables are usually sparse, as different rows may have different column families with data. Moreover, since data can be updated, the timestamp keeps multiple versions of a cell, as it changes over time. Different versions of a cell are stored in decreasing timestamp order, with the newest value in the first place, privileging current values over historical ones. The user can specify the number of versions that should be kept in the system, limiting the historical values to a certain period of time (HBase, 2018b).

When storing data, columns are often referenced with a `family:qualifier` pair, where the qualifier is any arbitrary array of bytes, corresponding to the name of the column in the column family. Figure 3-9 shows how a cell's history is recorded by HBase, tracking the changes undergone by the data as time passes.

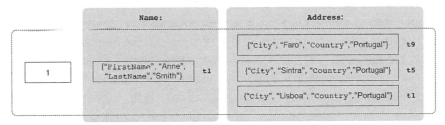

Figure 3-9. Timestamp in HBase tables.

Putting together the previous example, with the rows included in Figure 3-8 (table of students) and some versions of row 1, shown in Figure 3 9, two structures are physically created in HBase; one for storing all the rows belonging to the `Name` column family, and the other for storing all rows in the `Address` column family. The first table (Figure 3-10) includes two rows per record, one for storing the information about the `FirstName`, and the other for the `LastName`, since both components are included in the `Name` column family.

Row Key	Column Key	Timestamp	Cell Value
1	Name:FirstName	t1	Anne
1	Name:LastName	t1	Smith
2	Name:FirstName	t3	Bruce
2	Name:LastName	t3	Jones
3	Name:FirstName	t6	Mary
3	Name:LastName	t6	King
4	Name:FirstName	t10	Peter
4	Name:LastName	t10	Ralph

© FCA

Figure 3-10. HBase table for the `Name` column family.

In the second table (Figure 3-11) one of the records had several changes over time (record with the row key `1`). The table includes two rows per record at a given timestamp for storing the initial information about the `City` and the `Country`, plus a row per change, namely in the `City` column. The record with row key `1` presents changes at timestamps `t5` and `t9`.

In the previous example, the timestamp was represented by an abstract value to keep the example as simple as possible. Nevertheless, it is important to mention that the timestamp is a long integer value representing the number of seconds since midnight, January 1, 1970 UTC (Coordinated Universal Time), value known as the Unix time or Unix epoch (HBase, 2018b).

Having introduced the main concepts related to data models in HBase, the `hbase shell` is now used to create the `students` table with its two column families, `name` and `address`. For this table, the command line is also used to fill the sample records in the table. Using the `hbase shell`, the following code lines show how the students table is created (`create` command) with the two column families. For the `address` column family, the number of versions is set to 5, meaning that the cells' history for this column family will keep up to five different values at most. After creating the table, the content of HBase is listed (`list` command), showing the available tables, and data about two rows is inserted in the table using the `put` command.

Row Key	Column Key	Timestamp	Cell Value
1	Address:City	t1	Lisboa
1	Address:Country	t1	Portugal
2	Address:City	t3	Madrid
2	Address:Country	t3	Spain
1	Address:City	t5	Sintra
3	Address:City	t6	Paris
3	Address:Country	t6	France
1	Address:City	t9	Faro
4	Address:City	t10	Coimbra
4	Address:Country	t10	Portugal

Figure 3-11. HBase table for the `Address` column family.

```
$ hbase shell
hbase> create 'students', 'name',  {NAME => 'address', VERSIONS => 5}
0 row(s) in 1.2310 seconds
=> Hbase::Table - students
hbase> list
TABLE
students
1 row(s) in 0.0140 seconds
=> ["students"]
hbase> put 'students', '1', 'name:firstname', 'Anne'
hbase> put 'students', '1', 'name:lastname', 'Smith'
hbase> put 'students', '1', 'address:city', 'Lisboa'
hbase> put 'students', '1', 'address:country', 'Portugal'
hbase> put 'students', '2', 'name:firstname', 'Bruce'
hbase> put 'students', '2', 'name:lastname', 'Jones'
hbase> put 'students', '2', 'address:city', 'Madrid'
hbase> put 'students', '2', 'address:country', 'Spain'
```

At this moment, it is possible to inspect the table and see the inserted data. To do that, the scan command is used, listing 8 lines; 4 for each available record. In the obtained result, the sort command applied by HBase to the rows is shown, as the various columns of a record are sorted on disk by row key and column key, in a descending order by timestamp.

```
hbase> scan 'students'
ROW      COLUMN+CELL
 1        column=address:city, timestamp=1522750997363, value=Lisboa
 1        column=address:country, timestamp=1522750997474, value=Portugal
 1        column=name:firstname, timestamp=1522750997300, value=Anne
 1        column=name:lastname, timestamp=1522750997337, value=Smith
 2        column=address:city, timestamp=1522750997582, value=Madrid
 2        column=address:country, timestamp=1522750999662, value=Spain
 2        column=name:firstname, timestamp=1522750997517, value=Bruce
 2        column=name:lastname, timestamp=1522750997547, value=Jones
2 row(s) in 0.0320 seconds
```

The example continues updating the city of row key 1 and listing the table content to see the updated value as the current one.

```
hbase> put 'students', '1', 'address:city', 'Sintra'
hbase> scan 'students'
ROW      COLUMN+CELL
 1        column=address:city, timestamp=1522751761541, value=Sintra
 1        column=address:country, timestamp=1522750997474, value=Portugal
 1        column=name:firstname, timestamp=1522750997300, value=Anne
```

```
1       column=name:lastname, timestamp=1522750997337, value=Smith
2       column=address:city, timestamp=1522750997582, value=Madrid
2       column=address:country, timestamp=1522750999662, value=Spain
2       column=name:firstname, timestamp=1522750997517, value=Bruce
2       column=name:lastname, timestamp=1522750997547, value=Jones
2 row(s) in 0.1050 seconds
```

Now, the record with the row key 1 is fetched using the `get` command following two different approaches. The first one fetches the current value of the record, for both column families, while the second one fetches the two versions of the city already available in the table for the `address` column family.

```
hbase> get 'students', "1"
COLUMN                  CELL
 address:city            timestamp=1522751761541, value=Sintra
 address:country         timestamp=1522750997474, value=Portugal
 name:firstname          timestamp=1522750997300, value=Anne
 name:lastname           timestamp=1522750997337, value=Smith
4 row(s) in 0.0330 seconds
hbase> get 'students', "1", {COLUMN => 'address', VERSIONS => 2}
COLUMN                  CELL
 address:city            timestamp=1522751761541, value=Sintra
 address:city            timestamp=1522750997363, value=Lisboa
 address:country         timestamp=1522750997474, value=Portugal
3 row(s) in 0.0230 seconds
```

The remaining of the table is filled with the data previously shown in Figures 3-10 and 3-11, adding two more records (with row key 3 and row key 4), and with the additional update of the city for the student with the row key 1.

```
hbase> put 'students', '3', 'name:firstname', 'Mary'
hbase> put 'students', '3', 'name:lastname', 'King'
hbase> put 'students', '3', 'address:city', 'Paris'
hbase> put 'students', '3', 'address:country', 'France'
hbase> put 'students', '1', 'address:city', 'Faro'
hbase> put 'students', '4', 'name:firstname', 'Peter'
hbase> put 'students', '4', 'name:lastname', 'Ralph'
hbase> put 'students', '4', 'address:city', 'Coimbra'
hbase> put 'students', '4', 'address:country', 'Portugal'
```

To conclude the example, the content of the table is shown with the `scan` command, and the three versions of the city for the student to which it was assigned the row key 1 are shown with the `get` command. The students table is disabled afterwards with the `disable` command so it can be removed from HBase with the `drop` command.

```
▌ hbase> scan 'students'
▌ ROW      COLUMN+CELL
   1        column=address:city, timestamp=1522752674191, value=Faro
   1        column=address:country, timestamp=1522750997474, value=Portugal
   1        column=name:firstname, timestamp=1522750997300, value=Anne
   1        column=name:lastname, timestamp=1522750997337, value=Smith
   2        column=address:city, timestamp=1522750997582, value=Madrid
   2        column=address:country, timestamp=1522750999662, value=Spain
   2        column=name:firstname, timestamp=1522750997517, value=Bruce
   2        column=name:lastname, timestamp=1522750997547, value=Jones
   3        column=address:city, timestamp=1522752674146, value=Paris
   3        column=address:country, timestamp=1522752674168, value=France
   3        column=name:firstname, timestamp=1522752674075, value=Mary
   3        column=name:lastname, timestamp=1522752674122, value=King
   4        column=address:city, timestamp=1522752674264, value=Coimbra
   4        column=address:country, timestamp=1522752675368, value=Portugal
   4        column=name:firstname, timestamp=1522752674217, value=Peter
   4        column=name:lastname, timestamp=1522752674239, value=Ralph
▌ 4 row(s) in 0.0420 seconds
▌ hbase> get 'students', "1", {COLUMN => 'address:city', VERSIONS => 5}
▌ COLUMN            CELL
   address:city      timestamp=1522752674191, value=Faro
   address:city      timestamp=1522751761541, value=Sintra
   address:city      timestamp=1522750997363, value=Lisboa
▌ 3 row(s) in 0.0350 seconds
▌ hbase> disable 'students'
▌ hbase> drop 'students'
```

3.2.2.3. From Relational Models to HBase Data Models

After presenting the characteristics of column-oriented data models, and HBase in more detail, this subsection highlights how the transformation from a relational data model to a columnar data format, based on HBase, can be achieved.

As highlighted previously in this book, in a Big Data context the ability to collect, process, and store data increases with the use of NoSQL databases. These databases are characterized by being schema-free, allowing the storage of huge amounts of data without worrying excessively about the data structure. Issues usually emerge later on a schema-on-read approach, in which data is parsed, formatted, and cleansed at runtime. Although it avoids certain issues at the beginning of the collection phase, this increases the number of tasks later, when there is the need to develop specific applications to store, process, and analyze the data. At some point, these schema-free repositories need to be transformed into a structured data model that allows users to manipulate the data.

© FCA

In order to transform a relational data model into a columnar data model, namely for HBase, it is necessary to consider that a column-oriented database is composed of a set of tables integrating rows, but organized by groups of columns usually referred to as column families. This organization partitions the data vertically. Thus, each column family may contain a variable number of columns, even allowing missing columns between rows within the same table (HBase, 2018b).

Given the former contexualization, the following definitions formalize the concepts of relational data model and columnar data model.

Definition 1 – A Relational Data Model (RDM), $RDM = (T, A)$, includes a set of tables $T = \{T_1, T_2, ..., T_n\}$ and the corresponding attributes, $A = \{A_1, A_2, ..., A_n\}$, where A_1 is the set of attributes for table T_1, $A_1 = \{A_1^1, A_1^2, ..., A_1^k\}$. Tables in a database schema are linked with each other through relationships, with cardinalities of type 1:n, m:n or 1:1, with the optional 0 when needed. Each table includes an attribute that represents the primary key (PK) and may include one or more attributes representing foreign keys (FKs), linking this table to other tables in the database.

Definition 2 – A Columnar Data Model (CDM), $CDM = (T, CF)$, includes a set of tables $T = \{T_1, T_2, ..., T_n\}$. Each table integrates a key and a set of column families, as $T_i = \{key_i, CF_1, ..., CF_m\}$. Each column family integrates a set of columns representing the atomic values to be stored, $CF_j = \{C_j^1, ..., C_j^k\}$.

Considering these definitions, the following rules (C. Costa & Santos, 2017a; Santos & Costa, 2016a) can be adopted for transforming a RDM into a CDM. This set of rules can also be used to make the necessary transformations for other NoSQL databases based on the columnar format, such as Cassandra. This is possible as long as the necessary adaptations are performed, since the organization of the columns, or the way the key attribute of each table is defined, may vary.

Rule CDM.1 – Identification of column families. Identifying the column families of a CDM requires a two-step approach (see rules CDM.1.1 and CDM.1.2).

Rule CDM.1.1 – Identification of descriptive column families. All tables that do not include any FK in the RDM represent descriptive column families in the CDM, since these "core tables" store data associated with the main entities in the application domain, like `Students`, `Teachers`, `Facilities`, among others. These are the core descriptive column families that may also help to identify the complementary descriptive column families. These complementary column families are derived from tables that are linked to core tables through a 1:n relationship between the core table and the complementary table, and are usually split during the normalization process. Usually, these tables are used to complement the description of the core entities or are tables that need those core entities to wholly describe themselves. As an example, in the relationship a `Department` has many `Teachers`, `Department` is the core table and `Teachers` the complementary table, since only with `Department` it is possible to know the teachers' department. These complementary tables remain mostly static over time, since the set of `Teachers` of a `Department` does not change daily, for instance. This represents a type of relationship that is different from a `Student has many Evaluations`, for example, since, over time, any given `Student` will undergo many `Evaluations`. Rule CDM.1.1 does not handle this type of relationship, in which the entities are used to feed business processes instead of describing the main entities of the application domain. After identifying the descriptive column families, both core and complementary, their columns must also be identified. The columns of a descriptive column family are constituted by the set of non-key attributes (excluding primary or foreign keys) of the corresponding tables in the RDM.

Rule CDM.1.2 – Identification of analytical column families. All tables present in the RDM not identified by rule CDM.1.1 as descriptive column families give origin to analytical column families, integrating the day-by-day activities of an application domain or the main changes in the characterization of the core entities of the application domain. These analytical column families will be stored, processed, and analyzed taking into account the various descriptive column families. Rule CDM.1.2 excludes tables from the RDM that lack any attributes beside keys (PK or FK). The columns of analytical column families are constituted by the set of non-key attributes (excluding PKs or FKs) of the corresponding tables in the RDM.

Rule CDM.2 – Identification of tables. Two types of tables are proposed for a CDM, namely descriptive and analytical tables.

Rule CDM.2.1 – Descriptive tables. Descriptive tables are those that support specific data management tasks within an operational system. Each descriptive column family identified by rule CDM.1.1 gives origin to a descriptive table.

Rule CDM.2.2 – Analytical tables. To identify the set of analytical tables, the data workflows present in the RDM must be identified first. To do so, all tables present in the RDM which have not yet been identified by rule CDM.1.1 as descriptive column families start a data workflow following their corresponding $n{:}1$ relationships, and all subsequent $n{:}1$ relationships. The data workflow halts when no more $n{:}1$ relationships can be found; it follows that it is possible to rejoin coherent pieces of information that are related to each other but were split in the RDM during the normalization process. This means, for example, that a specific table in the RDM starts a data workflow and two or more tables without any FK end it. All identified workflows give origin to analytical tables.

Rule CDM.3 – Integration of column families into tables. A specific table integrates a key, a very important component of a table in a CDM, and a set of column families, which may vary depending on the table's type.

Rule CDM.3.1 – Column families of descriptive tables. A descriptive table derived from a core descriptive column family will incorporate this core descriptive column family as its unique column family. A descriptive table derived from a complementary descriptive column family will include as column families the complementary and the core descriptive column families to which it is associated. In case there are multiple 1:n relationships between various core descriptive column families and a complementary descriptive column family, the related set of column families will be integrated in the descriptive table.

Rule CDM.3.2 – Column families of analytical tables. An analytical table includes as column families the descriptive column families and, if applicable, the analytical column families included in the data workflow that gave origin to the specific analytical table according to rule CDM.2.2.

Rule CDM.4 – Definition of the tables' key. A table's key should be able to ensure an adequate performance throughout all read and write access patterns from applications. The key represents a set of one or more (concatenated) attributes that can form a natural key that properly identifies each row in the CDM. This key must serve the applications' get, scan and put patterns, keeping them as short as possible, while remaining potentially adequate for access patterns (HBase, 2018b). The order in which the attributes are concatenated plays a relevant role in the key's design, since HBase stores keys in a sorted order (Khurana, 2012).

To illustrate the usefulness of the proposed set of rules, the sample data model available in Hewitt (2010) is used as an example, where a set of HBase tables that fully complement each other while providing support for business applications is derived. The RDM used as starting point is based on the entity-relationship diagram shown in Figure 3-12, which integrates eight tables with several attributes, including PKs and FKs, as also shown in Figure 3-13.

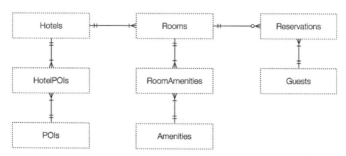

Figure 3-12. Entity-Relationship Diagram for the Hotel example. Adapted from Hewitt (2010).

In the above example, starting with rule CDM.1, specifically rule CDM.1.1, two types of descriptive column families are identified. The core descriptive column families are $Hotels_{CF}$, $POIs_{CF}$, $Guests_{CF}$ and $Amenities_{CF}$ because these tables only accept relationships with cardinality 1 and do not integrate any FK. A complementary descriptive column family, associated with the Rooms table, is then identified, giving origin to the $Rooms_{CF}$. A table in the model gives origin to an analytical column family (rule CDM.1.2), namely $Reservations_{CF}$. In this model, HotelPOIs and RoomAmenities are not identified as analytical column families because they do not include any attribute besides keys. The attributes of the identified column families are the attributes of the corresponding tables in the RDM, without the keys (either PKs or FKs).

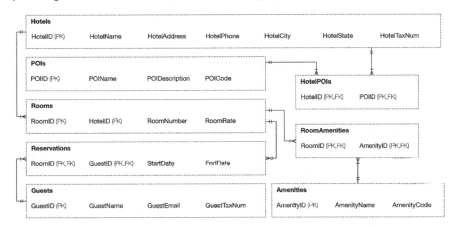

Figure 3-13. Relational data schema for the Hotel example. Adapted from Hewitt (2010) and C. Costa & Santos (2017a).

Regarding rule CDM.2, the descriptive tables are $Hotels_T$, $Rooms_T$, $POIs_T$, $Guests_T$, and $Amenities_T$ associated with the identified descriptive column families (rule CDM.2.1); to identify the analytical tables the data workflows in the RDM (rule CDM.2.2) first need to be identified. Using all the remaining tables in the RDM that were not identified as descriptive column families, HotelPOIs, RoomAmenities, and Reservations, three data workflows are identified (Figure 3-14) following the $n{:}1$ relationships starting in these tables and following this type of relationship until no other $n{:}1$ relationships are found. With this procedure, the three workflows generate three analytical tables, now called $Reservations_T$, $RoomAmenities_T$, and $HotelPOIs_T$.

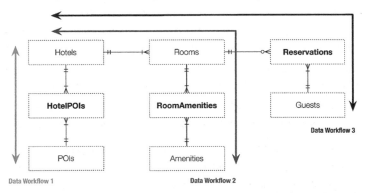

Figure 3-14. Data workflows for the Hotel example. Adapted from C. Costa & Santos (2017a).

After identifying the column families and the various tables, it is now necessary to assign the respective column families to each table (rule CDM.3.1). Each descriptive table integrates the core column family, or a set of column families including the core and the complementary descriptive column families, from which it was derived. $Rooms_T$ was derived from the complementary descriptive column family $Rooms_{CF}$; besides this column family, the HBase table also needs to include the column family associated with the $Hotels_{CF}$, because it is the core descriptive column family. $Hotels_T$, $POIs_T$, $Guests_T$, and $Amenities_T$ include the corresponding core column families, namely $Hotels_{CF}$, $POIs_{CF}$, $Guests_{CF}$, and $Amenities_{CF}$.

For the analytical tables (rule CDM.3.2), $Reservations_T$ integrates four column families, $Hotels_{CF}$, $Rooms_{CF}$, $Reservations_{CF}$, and $Guests_{CF}$; $RoomAmenities_T$ integrates three column families, $Hotels_{CF}$, $Rooms_{CF}$, and $Amenities_{CF}$; and $HotelPOIs_T$ integrates two column families, $Hotels_{CF}$ and $POIs_{CF}$.

After following the proposed rules, the identified CDM for HBase includes five descriptive tables and three analytical tables, as shown in Figure 3-15, which also shows the proposed keys (rule CDM.4) and columns.

The resulting data model allows managing all the data in the descriptive tables, for example, inserting/updating/deleting hotels, rooms, guests, amenities and POIs, as well as updating and including new facts generated by the organization's daily activities, such as new reservations, new amenities for the available rooms, or new POIs near the hotels' surroundings. Besides incorporating these new facts, it is possible to query the available data, answering questions from operational or tactical managers, thus supporting decision-making processes.

When the above model is compared to the one presented in Hewitt (2010), we can see that although both models have almost the same number of tables (7 vs. 8), these are organized in different ways. Whereas the model here presented includes descriptive tables for inserting new POIs or amenities, the model proposed in Hewitt (2010) only includes base tables for hotels, guests, and rooms. Reservations and POIs by hotel are organized in a similar way in both models. The main difference between the model we obtained and the one proposed by Hewitt (2010) is that in our model all the available data can be manipulated; whereas in the latter, some tables are included to provide answers to specific queries, such as available rooms or hotels sorted by city. Nevertheless, both data models can answer the same questions, as they include the same base columns.

At this point, it is important to mention that while the data model here obtained is adequate for HBase, the one proposed in Hewitt (2010) is best suited for Cassandra, noting the many different ways of organizing a column-oriented data schema.

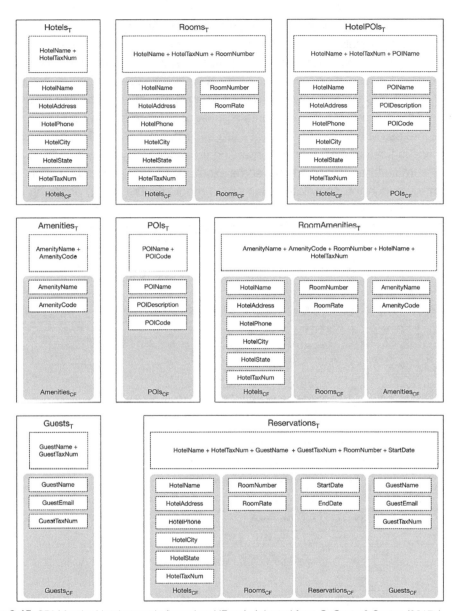

Figure 3-15. CDM for the Hotel example (based on HBase). Adapted from C. Costa & Santos (2017a).

When we think about a context of implementation where different queries can be used to analyze the available data, it is important to identify different analytical attributes that can be added to the analytical tables to provide a richer data model. Regarding the three previous analytical tables, $HotelPOIs_T$, $Reservations_T$, and $RoomAmenities_T$, there are different column families, either descriptive or analytical, which can be combined to derive new attributes with enhanced analytical capabilities. Using $HotelPOIs_T$ as example, it would be interesting to analyze whether the demand for a specific Hotel correlates, or not, with its proximity to a particular POI. For that, the average distance between each Hotel and a given POI can be derived during the ETL process, thus making that information available to the users. In the case of `Reservations`$_T$, the start and end dates can be used to derive the number of nights a guest stayed in a specific hotel; this information can be used later to calculate the average number of nights a guest usually stays. For `RoomAmenities`$_T$, it would be interesting to know the cost of extra amenities, when applicable, which allows the identification of the guests' profile regarding the money they are willing to pay, or not, for those extra features. These are just some examples of the derived data attributes that can be integrated in the analytical tables of the model (Figure 3-16), enriching the user analytical context.

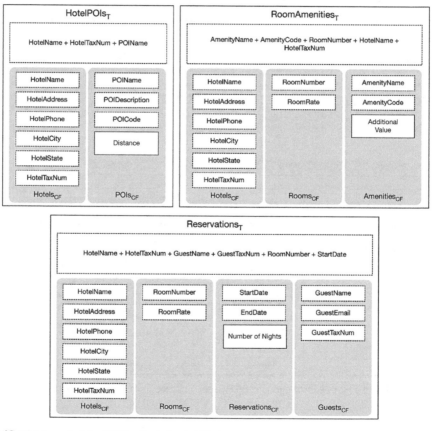

Figure 3-16. Enrichment of the CDM with derived attributes.

After showing how the proposed rules can be applied, a similar exercise for transforming another RDM into a CDM is now presented. The Transaction Processing Council (TPC) Benchmark H (TPC-H) is a decision support benchmark that includes a set of business oriented *ad hoc* queries and concurrent data modifications (TPC, 2017b). It is normally used for benchmarking decision support systems that process large volumes of data. The resulting CDM can be used for benchmarking processing technologies that store data in denormalized tables.

The data model of the TPC-H benchmark is shown in Figure 3-17. This data model includes eight transactional tables that contain sales, customers, and suppliers, thus modelling a business that manages, sells, and distributes products.

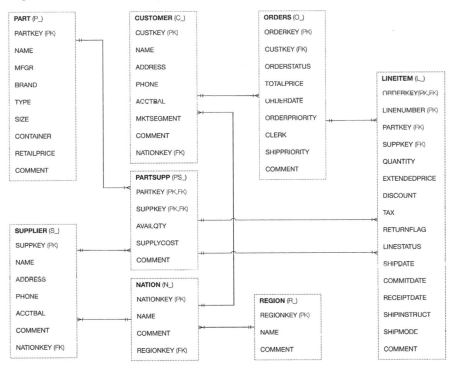

Figure 3-17. RDM for the TPC-H data model. Adapted from TPC (2017b) and C. Costa & Santos (2017a).

Following the proposed rules, and starting with rule CDM.1, when rule CDM.1.1 is applied, two types of descriptive column families are identified. The core descriptive column families are $Region_{CF}$ and $Part_{CF}$, because these tables only accept relationships with cardinality 1 and do not integrate any FK. Three complementary descriptive column families are identified, namely $Nation_{CF}$, $Supplier_{CF}$, and $Customer_{CF}$. Three other tables give origin to analytical column families (rule CDM.1.2), namely $PartSupp_{CF}$, $Orders_{CF}$, and $LineItem_{CF}$. The attributes of the identified column families are the same attributes of the corresponding tables in the RDM, except for the PKs or FKs.

Regarding rule CDM.2, the descriptive tables are $Region_T$, $Part_T$, $Nation_T$, $Supplier_T$, and $Customer_T$, as they are associated with the identified descriptive column families (rule CDM.2.1); but to identify the analytical tables, the data workflows of the RDM (rule CDM.2.2) need to be identified first.

To identify the three workflows in Figure 3-18 we used all the tables in the RDM that were not previously identified as descriptive column families (`PartSupp`, `Orders`, and `LineItem`). We did so by using the $n:1$ relationships starting in these tables and following all the related entities until no more $n:1$ relationships were found. Following this procedure, the three data workflows generate three analytical tables, namely, $PartSupp_T$, $Orders_T$, and $LineItem_T$.

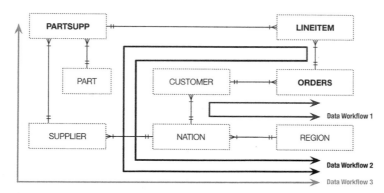

Figure 3-18. Data workflows for the TPC-H dataset. Adapted from C. Costa & Santos (2017a).

Having identified the column families and tables, it is now necessary to assign the respective column families to each table. Each descriptive table (rule CDM.3.1) integrates a core column family, or a set of column families including the core and the complementary descriptive column families from which it was derived.

$Customer_T$ was derived from the complementary descriptive column family $Customer_{CF}$. Besides this column family, it also needs to include the $Nation_{CF}$ complementary descriptive column family and the $Region_{CF}$ core descriptive column family. $Region_T$ and $Part_T$ include the corresponding core column families, $Region_{CF}$ and $Part_{CF}$, respectively. $Nation_T$ includes the core $Region_{CF}$ and the complementary $Nation_{CF}$; $Supplier_T$ includes $Nation_{CF}$ and $Supplier_{CF}$ as complementary column families and $Region_{CF}$ as core column family.

For the analytical tables (rule CDM.3.2), $PartSupp_T$ includes five column families, $PartSupp_{CF}$, $Part_{CF}$, $Supplier_{CF}$, $Nation_{CF}$, and $Region_{CF}$; $Orders_T$ integrates four column families, $Orders_{CF}$, $Customer_{CF}$, $Nation_{CF}$, and $Region_{CF}$; and $LineItem_T$ integrates the eight column families, achieving a complete denormalization of the relational schema: $LineItem_{CF}$, $PartSupp_{CF}$, $Part_{CF}$, $Supplier_{CF}$, $Orders_{CF}$, $Customer_{CF}$, $Nation_{CF}$, and $Region_{CF}$. Moreover, it is worth mentioning that $Nation_{CF}$ and $Region_{CF}$ have been denormalized twice, since it was necessary to specify the suplier's and the customer's nation and region.

After following the proposed rules, the identified CDM for HBase includes five descriptive tables, as shown in Figure 3-19, along with the proposed keys (rule CDM.4) and columns. Regarding the analytical tables, and also showing the proposed keys (rule CDM.4) and columns, Figure 3-20 highlights the three resulting analytical tables and the logical aggregation of column families to verify the various denormalizations of Nation$_{CF}$ and Region$_{CF}$.

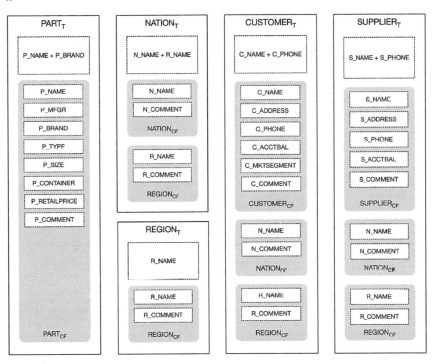

Figure 3-19. CDM with the descriptive tables for the TPC-H data model. Adapted from C. Costa & Santos (2017a).

After following the proposed rules for transforming a RDM into a CDM, several tables were obtained, which serve different purposes in terms of data management. Some tables have a more operational role, maintaining information about the main entities of an application domain, whereas other tables present an analytical profile, allowing the user to answer specific questions concerning the business processes of an organization. It is up to the data engineers to decide which tables must be implemented, taking into consideration the data management and analytical tasks that must be supported. In the case of the analytical tables PartSupp$_T$ and Orders$_T$, they are fully contained in LineItem$_T$, therefore, all queries can be answered by the latter, making the implementation of the former tables redundant. The main difference between the tables is their size, a factor that may influence performance when responding to queries.

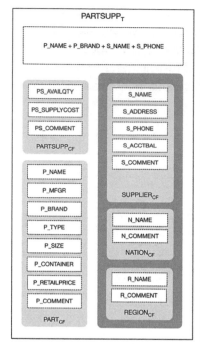
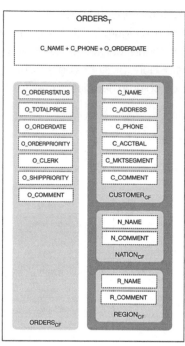
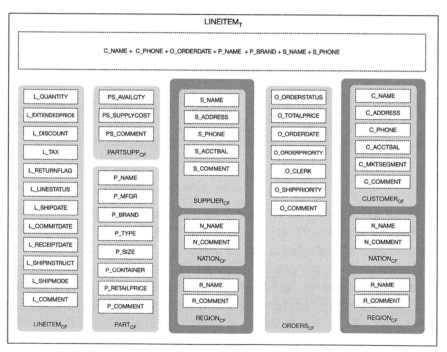

Figure 3-20. CDM with the analytical tables for the TPC-H data model. Adapted from C. Costa & Santos (2017a).

Another aspect that depends on the business context in which these tables are going to be used is the definition of their keys. It is important to note that the keys used here are merely examples of how keys can be composed and defined for each table. The keys can be more appropriate for inserting or searching data, depending on the attributes and concatenation order between them.

3.2.3. Document-oriented Databases

This subsection provides an overview of the document stores and describes MongoDB in more detail, with some examples of its data model and functionalities.

3.2.3.1. Overview

Document stores, or document databases, are similar to key-value stores, as they also require a key to access the data. However, in their case, the value is stored in a document that contains structured or semi-structured data, which can be queried with appropriate tools (Figure 3-21). A document usually includes an identifier and a nested structure. The fields included in the nested structure can hold a wide variety of data types, such as integers, floats, strings, spatial coordinates, among others. One major difference between document stores and key-value stores is that in the former it is possible to query both the content of the document and the key (Sadalage & Fowler, 2012).

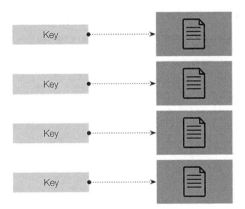

Figure 3-21. Document stores.

Documents have a flexible structure (schema-free organization of the data), and each document can support a different data model, as different attributes can be added when needed. This means the rows in a given dataset do not need to have the same data structure, with the same columns. Furthermore, the type of values in the columns can vary from row to row. Columns can include a cell value or an array referring to more than

© FCA

one value for that row/column. Besides this, rows can have nested structures that embed objects inside other existing objects.

Documents are often stored in XML, JSON, or BSON (Binary JSON) formats, although they can also store `.doc`, `.docx`, `.pdf`, and many other file types. Instead of including columns with names and their respective data types, as is the case in a RDBMS, documents in a document store contain the description of the data type (attribute) and the corresponding values.

Figure 3-22 exemplifies how a JSON oriented-format is used for storing the values of a collection of documents that include the `student_number` data type and the values for a `student_name`, `degree_description`, as well as a nested data type called `curricular_units` detailing the students' curricular units, namely the `unit_name`, `unit_year`, and `unit_grade`. Since data schemas are very flexible, it is possible to have variations in the data structure, such as the `unit_class` data type for some curricular units. The rationale behind a document-oriented model is that all data about a specific row can be stored within the same document, avoiding the need to establish relationships with other documents in the database.

```
{
    'student_number': 12345,

    'student_name': {'Peter Ralph'},

    'degree_name': {'Applied Mathematics'},

    'curricular_units': [

        {'unit_name': 'Introduction to Algebra',
         'unit_year': 1,
         'unit_grade': 16},

        {'unit_name': 'Integral Calculus',
         'unit_year': 1,
         'unit_grade': 18},

        {'unit_name': 'Finite Mathematics',
         'unit_year': 1,
         'unit_grade': 17,
         'unit_class': 'Not mandatory'} ]

}
```

Figure 3-22. Example of a JSON document.

As shown in the previous example, documents describe themselves through the stored data and can contain nested structures that give them a hierarchical data structure which is useful for establishing explicit semantic relationships among the data. Regarding the terminology, a specific database instance can be seen as a collection of documents, each one corresponding to a row or entry in that database.

Document databases scale horizontally and have fast write performance, since write availability is prioritized over data consistency. Writing will remain fast even when a failure (in the available hardware or network) delays data replication and, consequently, data consistency over the cluster. Aditionally, fast query performance is achieved through appropriate query engines (including engines written in Java, JavaScript, or Python, for example) and database's indexing mechanisms, which generally ensure fast and efficient querying capabilities.

3.2.3.2. MongoDB

MongoDB (2018) is one of the best known document databases. It provides high performance, high availability, and auto-scaling. Each record in MongoDB is a document where the value field may include other documents, arrays, or arrays of documents. Documents have flexible data schemas, which are dynamically adjusted depending on the data requirements.

MongoDB (2018) includes a query language that supports Create, Read, Update, and Delete (CRUD) operations, as well as data aggregations (based on the concept of pipelines), text search (looking into strings' content) and geospatial queries (for performing spatial queries on a collection of documents containing geospatial data).

This type of database also includes a replication facility (the replica set) that provides automatic failover and data redundancy. The replica set is a group of MongoDB servers that include the same data set, thus increasing data availability. MongoDB stores data using the BSON format.

Documents are the working unit of data in MongoDB, similar to a row in a RDBMS, but complex enough to allow more flexibility when storing data. Documents are grouped in collections. Each document in a collection has a unique key, _id. In case no _id key is available when a document is inserted, the system will automatically generate one for it (Chodorow, 2013). When the _id is automatically generated, it follows the ObjectId default type. This type is designed to be lightweight and easy to generate in a globally unique way across different machines. Since MongoDB is a distributed database, the approach that is used to generate the key avoids the time-consuming task of synchronizing keys. The ObjectId uses 12 bytes to store a timestamp (a 4-byte value representing seconds in Unix time), as well as information about the machine (a 3-byte machine identifier), the process identifier of the ObjectId-generating process (a 2-byte process identifier) and an incrementing counter (a 3-byte counter) (Chodorow, 2013), as shown in Figure 3-23.

0	1	2	3	4	5	6	7	8	9	10	11
Timestamp				Machine			Process		Increment		

Figure 3-23. Structure of a document key in MongoDB. Adapted from Chodorow (2013).

To illustrate some of the functionalities of MongoDB and its query language, some examples are now shown. These include creating a database and storing a set of documents, which are later queried to answer specific questions. In the example database, the `students` collection is stored, along with four documents (Figure 3-24).

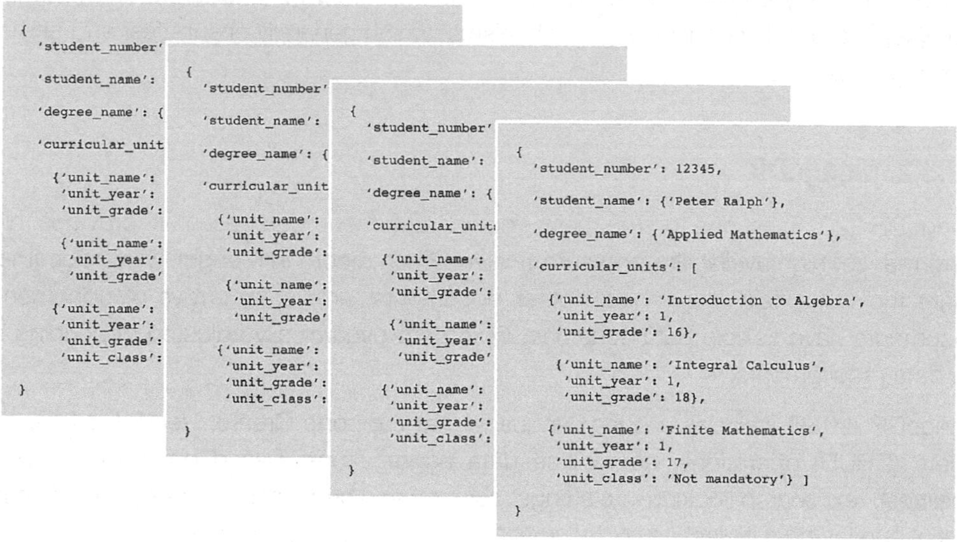

Figure 3-24. A collection of documents.

The MongoDB shell can be invoked using the `mongo` command and the database can be selected with the `use` command. If the database does not exist, MongoDB will create one when data is first stored. In the following example, the database is created along with the `students` collection when documents are inserted. In the command line, `db` points to the current database, and the `insertOne` command is used to create one document at a time. In this example, the student number is filled in the `_id` field, which also serves as the document key. This number must be unique to prevent errors when inserting data in the collection.

```
$ mongo
mongo> use example
mongo> db.students.insertOne(
    {_id:12345,
    student_name:"Peter Ralph",
    degree_name:"Applied Mathematics",
    curricular_units:[
        {unit_name:"Introduction to Algebra",
        unit_year:1,
        unit_grade:16},
        {unit_name:"Integral Calculus",
        unit_year:1,
```

```
        unit_grade:18},
        {unit_name:"Finite Mathematics",
        unit_year:1,
        unit_grade:17,
        unit_class:"Not mandatory"}
        ]
    })
```

For inserting several records in a specific collection, the `insertMany` command must be used. In the example, three additional students are added to the collection.

```
mongo> db.students.insertMany(
    [
        {_id:67890,
        student_name:"Anne Smith",
        degree_name:"Applied Mathematics",
        curricular_units:[
            {unit_name:"Integral Calculus",
            unit_year:1,
            unit_grade:12},
            {unit_name:"Introduction to Algebra",
            unit_year:1,
            unit_grade:14}
            ]},
        {_id:24680,
        student_name:"Bruce Jones",
        degree_name:"Physics",
        curricular_units:[
            {unit_name:"Experimental Physics",
            unit_year:2,
            unit_grade:17},
            {unit_name:"Quantum Physics",
            unit_year:3,
            unit_grade:20}
            ]},
        {_id:13579,
        student_name:"Mary King",
        degree_name:"Physics",
        curricular_units:[
            {unit_name:"Experimental Physics",
            unit_year:2,
            unit_grade:17},
            {unit_name:"Quantum Physics",
            unit_year:3,
            unit_grade:20}
            ]}
    ])
```

© FCA

For querying a collection, the find command must be used, with or without conditions. Without conditions, it returns all the documents available in the collection. For specifying conditions, different matching strategies can be followed:

- A specific value for a field (in the current example, for selecting all students in the degree with the "Physics" name);

- An exact document (equality matches), where the whole document needs to match exactly all the values embedded in a specific field, including the field order (in the example, for selecting all students with the "Quantum Physics" curricular unit from year "3" and a grade of "20");

- A field in an embedded document (where the field unit_name, embedded in the curricular_unit, must be equal to the "Introduction to Algebra" string).

```
mongo> db.students.find({})
mongo> db.students.find({degree_name: "Physics"})
mongo> db.students.find({curricular_units:
        {unit_name:"Quantum Physics",
        unit_year:3,
        unit_grade:20}})
mongo> db.students.find({"curricular_units.unit_name":"Introduction to Algebra"})
```

Variations of these queries can be performed to obtain a specific element from an array or even to obtain an exact match of all the elements a given array should contain. For specific examples, please refer to the MongoDB documentation (MongoDB, 2018).

Another querying approach relies on operators such as greater than ($gt), less than ($lt), the logical AND (combining several conditions separated by commas), and the logical OR ($or). In the following examples, the collection is queried to find all students with grades higher than 18 in any curricular unit; with grades lower than 14 in any curricular unit; information about a student called Bruce Jones enrolled in the Physics degree; and, finally, all students from the Physics or the Mechanics degrees.

```
mongo> db.students.find({"curricular_units.unit_grade":{$gt:18}})
mongo> db.students.find({"curricular_units.unit_grade":{$lt:14}})
mongo> db.students.find({degree_name: "Physics", student_name: "Bruce Jones"})
mongo> db.students.find({$or:[{degree_name: "Physics"},{degree_name: "Mechanics"}]})
```

For updating data, the updateOne or updateMany commands can be used. Using the updateMany, the following code updates the degree name, from Physics to Physics Engineering.

```
mongo> db.students.updateMany(
        {degree_name: "Physics"},
        {$set:{degree_name: "Physics Engineering"}})
```

Similar commands may be used to delete data, e.g., `deleteOne` or `deleteMany`. In the example below, `deleteOne` will erase the student named `Mary King`.

```
mongo> db.students.deleteOne({student_name: "Mary King"})
```

Besides being able to deal with different records individually, MongoDB can perform several aggregations on data values. To do so, the aggregation pipeline may be applied. This is a framework that takes several documents as input and provides aggregated values as a result. This aggregation pipeline can be seen as an alternative to MapReduce tasks, in cases where the complexity of such tasks is not justified.

Using a simplified version of the `students` example, Figure 3-25 shows how the aggregate function includes a filter to select the students enrolled in the "`Applied Mathematics`" degree, grouping them by student number and calculating each student's average grades.

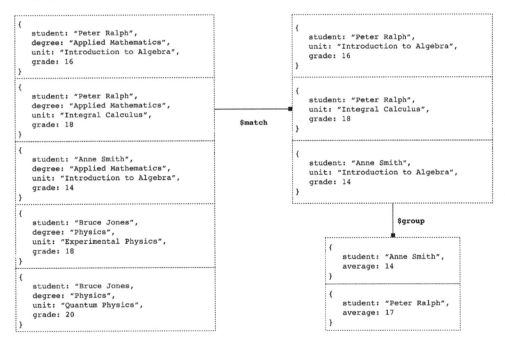

Figure 3-25. Aggregation pipeline of MongoDB.

For this example, the `grades` collection is created along with five documents. Since no `_id` is manually introduced, MongoDB assigns each document a unique `ObjectId`, as can be seen in the acknowledgment message returned by the system.

```
mongo> db.grades.insertMany([
          {student:"Peter Ralph", degree:"Applied Mathematics",
           unit:"Introduction to Algebra", grade:16},
```

```
            {student:"Peter Ralph", degree:"Applied Mathematics",
             unit:"Integral Calculus", grade:18},
            {student:"Anne Smith", degree:"Applied Mathematics",
             unit:"Introduction to Algebra", grade:14},
            {student:"Bruce Jones", degree:"Physics",
             unit:"Experimental Physics", grade:18},
            {student:"Bruce Jones", degree:"Physics",
             unit:"Quantum Physics", grade:20},
   ])
```

```
{"acknowledged": true,
 "insertedIds": [
      ObjectId("5c543325090362a9ee2bcd88"),
      ObjectId("5c543325090362a9ee2bcd89"),
      ObjectId("5c543325090362a9ee2bcd8a"),
      ObjectId("5c543325090362a9ee2bcd8b"),
      ObjectId("5c543325090362a9ee2bcd8c")
   ]}
```

To carry out this aggregation, a two-step process is followed using the `$match` operator, which selects all students enrolled in the specified degree. Afterwards the `$group` operator merges the documents corresponding to each student (i.e., those with the same `student_number`) and applies the `$avg` operator, which returns the requested averages (averages of the `unit_grade` values). This aggregation can be written in the MongoDB shell as follows:

```
mongo> db.grades.aggregate([
          {$match: {"degree": "Applied Mathematics"}},
          {$group: {_id: "$student", average: {$avg: "$grade"}}}
   ])
{"_id": "Anne Smith", "average": 14}
{"_id": "Peter Ralph", "average": 17}
```

Several accumulators can be used during the `$group` stage. Besides `$avg`, some of the most common include `$sum`, `$min`, and `$max`. For a detailed list, please visit the aggregation pipeline quick reference guide in (MongoDB, 2018). The following example shows how the `$sum` accumulator is used to verify the number of occurrences by grade, listed first in descending order according to the `count` variable, and then by ascending order according to the `count` variable.

```
db.grades.aggregate([
          {$group: {_id: "$grade", count: {$sum: 1}}},
          {$sort : {count: -1} }
   ])
```

```
{"_id": "18", "count": 2}
{"_id": "20", "count": 1}
{"_id": "14", "count": 1}
{"_id": "16", "count": 1}
db.grades.aggregate([
        {$group: {_id: "$grade", count: {$sum: 1}}},
        {$sort : {count: 1} }
  ])
{"_id": "20", "count": 1}
{"_id": "14", "count": 1}
{"_id": "16", "count": 1}
{"_id": "18", "count": 2}
```

Besides allowing the analysis of documents based on a more structured perspective of the data, MongoDB also supports query operators that allow searching text within strings. To do so, the `$text` operator is used along a text index on the fields that are going to be queried. Text indexes can be used on fields that contain string values or an array of string elements.

Using once more a simplified version of the `students` example, the following instructions create another students collection with four documents (called `tutors`), one for each student, with the key (`_id`), the `student_name` and the `degree_name`. Afterwards, the `createIndex` command applied to the `student_name` and `degree_name` fields creates a searching index for the text contained in those fields.

```
mongo> db.tutors.insertMany([
   {_id:12345, student_name:"Peter Ralph", degree_name:"Applied Mathematics"},
   {_id:67890, student_name:"Anne Smith", degree_name:"Applied Mathematics"},
   {_id:24680, student_name:"Bruce Jones", degree_name:"Physics"},
   {_id:13579, student_name:"Mary King", degree_name:"Physics"}])
mongo> db.tutors.createIndex({student_name: "text", degree_name: "text"})
```

Next, the `$text` operator can be used to search text in a collection. This can be done in different ways, depending on the type of search the user needs. For example, searching documents that contain any of the terms provided in a list, can be done as follows. In this case, the list includes three terms, "`Peter`", "`Applied`", and "`Mathematics`":

```
mongo> db.tutors.find({$text: {$search: "Peter Applied Mathematics"}})
```

The `$search` command can be used to query an exact phrase, however, it must be wrapped in double quotes, as shown in the following example, where the terms being looked for are "`Peter`" or "`Applied Mathematics`":

```
mongo> db.tutors.find({$text: {$search: "Peter \"Applied Mathematics\""}})
```

Excluding terms is also possible; in this case, explicitly indicating a term that cannot be found in a document, while simutaneously pointing to those that could be present. In the following example, the search covers all documents that include the terms "`Applied`" or "`Mathematics`", but not the term "`Peter`".

```
mongo> db.tutors.find({$text: {$search: "-Peter Applied Mathematics"}})
```

The last example related to MongoDB functionalities concerns the use of geospatial data. This type of data can be stored as a GeoJSON object or as legacy coordinate pairs. With this data, different spatial operators can be applied to carry out tasks such as calculating distances, areas, intersections, among many others, depending on the geometry of the objects. For legacy coordinate pairs, where data is stored as points in a two-dimensional plane, documents include coordinates represented by field values, and a `2d` index is required to calculate distances in an Euclidean plane or to perform other spatial queries on the data. In the following example, the `$near` operator is used to locate the points around a given location.

```
mongo> db.students_geo.insertOne(
    {_id:12345,
    student_name: "Peter Ralph",
    student_degree: "Applied Mathematics",
    student_location: [-8.5218107, 41.4128713]})
mongo> db.students_geo.ensureIndex({student_location: "2d"})
mongo> db.students_geo.find({student_location: {$near: [-8.5, 41]}})
```

For spherical surface calculations, legacy coordinates first need to be converted to GeoJSON data types. The GeoJSON format is used for representing geographic data structures, which supports geometry types such as `Point`, `LineString`, `Polygon`, `MultiPoint`, `MultiLineString`, and `MultiPolygon`. With GeoJSON, `2dsphere` supports spatial operations that can be performed on the `Point` data type, for example. In the following code, two documents are inserted in the `locations` collection, both represented by the `Point` data type and the respective coordinates pair. The `2dsphere` is used to index the data. For searching the data, `findOne` returns any document in the collection, whereas the `find` command looks for documents of the `Point` geometry that are near a specific location. The `$nearSphere` operator calculates distances using spherical geometry.

```
mongo> db.locations.insertMany([
    {name: "Peter Ralph",
     degree: "Applied Mathematics",
     location: {coordinates: [-73.856077,40.848447], type: "Point"}},
    {name: "Bruce Jones",
     degree: "Physics",
     location: {coordinates: [-73.961704,40.662942], type: "Point"}}
])
```

```
mongo> db.locations.createIndex({location: "2dsphere"})
mongo> db.locations.findOne()
mongo> db.locations.find({location:
   {$nearSphere:
       {$geometry: {type:"Point", coordinates: [-73.8, 40.7]}}}
})
```

Besides the `$nearSphere` operator, other geospatial query operators are available in MongoDB, such as `$geoWithin`. For a detailed list, please refer to the MongoDB documentation (MongoDB, 2018). For examples on other GeoJSON objects, such as `LineString`, `Polygon`, `MultiPoint`, `MultiLineString`, and `MultiPolygon`, please also refer to the MongoDB documentation (MongoDB, 2018). This subsection ends with an example that uses the `$centerSphere` command to identify all documents from the `locations` collection within a 20 km radius of the specified location. As `$centerSphere` uses radians for distances, the distance must be divided by the radius of the sphere (in this case, the Earth), which is approximately 6371 km.

```
mongo> db.locations.find({location:
   {$geoWithin:
       {$centerSphere: [[-73.8, 40.7], 20/6371]}}
})
```

3.2.4. Graph Databases

This subsection provides an overview of graph databases and presents Neo4j in more detail, with some examples of its data model and functionalities.

3.2.4.1. Overview

Graph databases are particularly useful for storing rows with a small number of constituting attributes but complex interconnections between them (Sadalage & Fowler, 2012). A graph data structure includes vertices (or nodes) connected by edges (or relationships); it allows to make queries about the data present in the nodes and the semantics of the connections explicitly stored in the edges.

Graph structures model different types of complex relationships between entities in the nodes; they are useful for making sense of a wide diversity of real world datasets (Robinson, Webber, & Eifrem, 2015). Figure 3-26 shows how the global data model behind a graph structure may be understood. The data rows exist both in nodes and in relationships. Data about the entities is stored in the nodes, whereas data about the relations or connections between the entities is stored in the relationships. Relationships thus serve to organize the entities. Both nodes and relationships have specific properties that characterize them.

© FCA

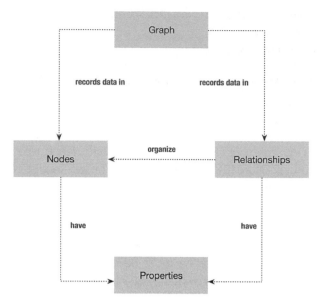

Figure 3-26. A graph data structure. Adapted from Robinson et al. (2015).

Figure 3-27 shows an example where two types of entities are modeled: curricular units and students. For the `Algebra` curricular unit, the `student` relationship links all students enrolled in the course, whereas the `is_student` relationship does the opposite, associating students with their units and the dates when these relationships began. For students, various records (different students) are linked to each other depending on the `knows` relationship and the `year` it began. In this context, nodes (representing the instances of each entity) have attributes (or properties) and so do the relationships (representing how these instances are related to each other).

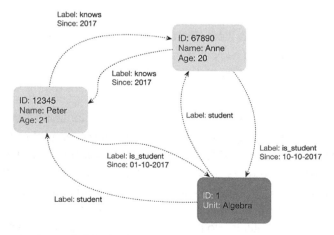

Figure 3-27. An example of a graph data structure.

As shown in the previous example, the most common graph model is the labeled property graph, which has the following characteristics (Robinson et al., 2015):

- Graphs contain nodes and relationships;

- Nodes contain properties represented by key-value pairs;

- Nodes can be labeled (using one or more labels);

- Relationships are labeled using a name and have a direction, from a start to an end node;

- Relationships contain properties that are also represented by key-value pairs.

Besides these characteristics, it is important to mention that nodes can have more than one relationship, since the same nodes can be linked through one or more relationships, and self-relationships can also exist, as illustrated in Figure 3-28.

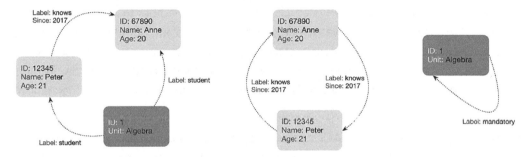

Figure 3-28. Different types of relationships in graphs. Adapted from Robinson et al. (2015).

Graph databases, or graph database management systems, support CRUD operations, being frequently used to implement operational data systems (OLTP), reason why they are optimized for transactional performance and transactional integrity. A graph database can use a native graph storage system optimized for storing and managing graphs or serialize the graph data into a relational database, an object-oriented database, or other data store. In terms of processing engine, a graph database can use an index-free adjacency. This means the connected nodes physically point to each other in the database (an explicit reference that significantly increases the performance of the system). Alternatively, the processing engine can expose a graph data model through CRUD operations, which have some impact on the performance of the system (Robinson et al., 2015).

The benefits of using a graph database emerge when there is the need to handle highly interconnected data, in a context where performance tends to remain constant, even when the dataset grows (since queries are usually located in a specific section of the graph). Such is not the case with relational databases, where performance costs from join-intensive queries increase as the datasets grow. In terms of data model, the dynamic structure of the graph allows it to evolve, hence it is possible to add new data requirements as soon as they become available in the application domain. New relationships, nodes, labels, or subgraphs can be added to a graph structure without damaging any existing applications or queries, since these are the features that characterize the schema-free nature of graph databases (Robinson et al., 2015).

3.2.4.2. Neo4j

Neo4j (2018) is a native graph storage system with a native graph processing engine that optimizes fast storage and management of highly interconnected data represented by nodes and relationships. Instead of using joins between tables, as happens in the relational data model, Neo4j navigates from one node to another in a linear fashion (a type of performance which, as was already mentioned, is not impacted by the size of the dataset, as is the case with relational databases). This performance is continuous, since most graph searches are carried out in the immediate surroundings of a node and this kind of database has an efficient and highly scalable memory management mechanism (Neo4j, 2018).

Neo4j is fully compatible with ACID properties; it uses transactions to guarantee data persistence in case of a system failure, and supports high-performance applications, being able to store hundreds of trillions of entities. Due to its high scalability, Neo4j can operate on clusters with dozens of machines, as opposed to hundreds or thousands, to handle such amounts of data (Neo4j, 2018).

The Cypher query language can be used for querying Neo4j. This declarative language matches patterns of nodes and relationships to extract or modify data in the graph, through CRUD operations and queries.

Cypher is a SQL-inspired language that uses ascii-art syntax to specify the data to be selected, created, inserted, updated, or deleted, without the need to specify how to do it. In the next example, the relationship `knows` between two students can be established as follows:

```
(student A): - [:knows] - > (student B)
```

In this syntax, nodes are surrounded by parentheses `(node)`. These nodes can be assigned to variables for later use, e.g., `(s:Student A)`.

The general structure of Cypher includes the MATCH statement for specifying what the user is looking for, the conditions of the search, in case they exist, and RETURN to mention which fields the user is expecting as a result. When no specific conditions are indicated, the syntax is as follows:

```
MATCH   (node:Label)
RETURN node.property
```

When using conditions in the WHERE clause, the syntax is as follows:

```
MATCH   (node1:Label1) - - > (node2:Label2)
WHERE   node1.property1 = {value}
RETURN node1.property1, node2.property2
```

In this syntax, the arrow is used to express relationships between nodes. Additional information about the relationship, such as a label, can be written between brackets inside the arrow: – [:knows] – >. In these cases, a variable can also be assigned for later reference in the WHERE and/or RETURN operations.

If the following example, a node is created as an instance of the Student entity, and the inserted value is returned to the user (Figure 3-29).

```
CREATE (peter:Student {name: 'Peter Ralph'})
RETURN peter
```

Peter Ralph

Figure 3-29. Node returned to the user in Neo4j.

To find the instance assigned to peter and simultaneously add a new relationship to a new node in the graph (Figure 3-30), the following code uses the MATCH, CREATE, and RETURN clauses.

```
MATCH (peter:Student {name: 'Peter Ralph'})
CREATE (peter) - [student:is_student] - >
                  (algebra:Unit {name:'Introduction to Algebra'})
RETURN peter, student, algebra
```

Figure 3-30. Establishing a relationship between two nodes in Neo4j.

The CREATE clause can add a single node or a more complex structure, such as simultaneously adding several instances to a graph. In this example, various friends are linked to a student using the knows relationship.

```
MATCH (peter:Student {name: 'Peter Ralph'})
FOREACH (friend in ['Anne Smith', 'Bruce Jones', 'Mary King']|
   CREATE (peter) - [:knows] - > (:Student {name:friend}))
```

Next, a select all operation can be performed to see all the records, showing all the nodes and relationships that have been already added to the graph (Figure 3-31). To do so, a global MATCH is done.

```
MATCH (n)
RETURN n
```

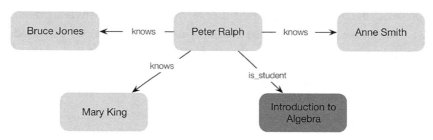

Figure 3-31. Selecting all the nodes and relationships in Neo4j.

To select a particular relationship (Figure 3-32), the MATCH clause must include the reference to that relationship. In this example, a variable is used to store the existing relationships as a result; this variable is later used by the RETURN clause.

```
MATCH (peter:Student {name: 'Peter Ralph'}) - [:knows] - > (peterFriends)
RETURN peter, peterFriends
```

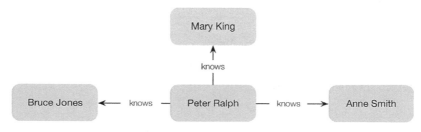

Figure 3-32. Selecting a specific relationship in Neo4j.

To create more complex relationships such as a hierarchy in the instances that belong to an entity, the existing nodes may be extended. In this example, the `Student` entity is extended to allow instances of the type `Tutor`, for students who are also tutors. Tutors are, in this case, former students of a `Unit`. For former students, a new relationship is established in the data model, `was_student`.

```
MATCH (algebra:Unit {name: 'Introduction to Algebra'})
MATCH (anne:Student {name: 'Anne Smith'})
CREATE (anne) -[:knows]- >
            (:Student:Tutor {name: 'Alex Wang'}) -[:was_student]- > (algebra)
```

At this point, the graph includes six nodes with data from three different entities: five students, one is also a tutor, and one unit. The data model also includes three different relationships between the nodes, namely, `knows`, `is_student`, and `was_student` (Figure 3-33).

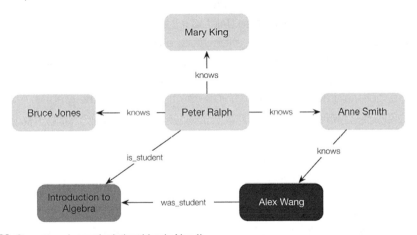

Figure 3-33. Current nodes and relationships in Neo4j.

Working with a graph means that existing paths are followed when querying the data. In the following example, the path between the nodes is followed to find out the shortest path (`shortestPath`) between a student and a tutor, using the `knows` relationships, considering no more than five edges between them (Figure 3-34).

```
MATCH (peter {name: 'Peter Ralph'})
MATCH (tutor) - [:was_student] - > (algebra:Unit {name: 'Introduction to
Algebra'})
MATCH path = shortestPath((peter) - [:knows*..5] - (tutor))
RETURN algebra, tutor, path
```

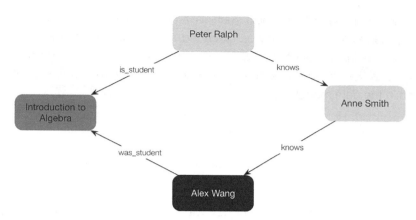

Figure 3-34. Finding the shortest path between two nodes in Neo4j.

To exemplify other CRUD operations, data can be deleted when looking for a specific node or relationship. A node cannot be deleted if it still maintains relationships with other nodes. To remove a relationship and, afterwards, a node, the following instructions can be used. In this example, the relationship `knows` between the two students is identified and deleted. Afterwards, the node of the student identified by the `mary` variable is deleted, since it no longer maintains any relationship with other nodes in the graph.

```
MATCH (n1 {name: 'Peter Ralph'}) - [r:knows] - > (n2 {name: 'Mary King'})
DELETE r
MATCH (mary:Student {name: 'Mary King'})
DELETE mary
```

After running all the previous instructions, the visual display of the resulting nodes and relationships is shown in Figure 3-35, using the Neo4j interface. The graph now includes five nodes with data from three different entities: four students, one of them is also a tutor, and one unit.

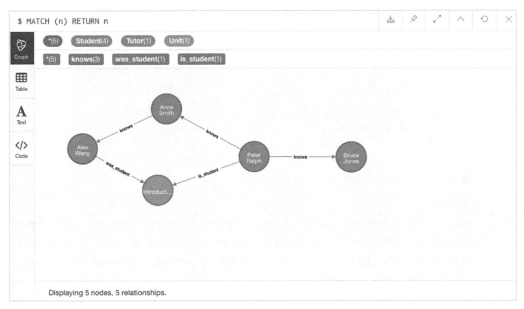

Figure 3-35. An example of a graph database in Neo4j graphical interface.

Neo4j is also suited for storing and querying geospatial data, either using a Cartesian reference system with a 2D coordinate system or the World Geodetic System 1984 (WGS84) for a spheroidal reference surface of the Earth.

In the following example, the distance between two Cartesian points is calculated; the returned value is 7.071:

```
WITH point({x:5,y:5,crs:'cartesian'}) AS p1,
     point({x:10,y:10,crs:'cartesian'}) AS p2
RETURN distance(p1,p2) AS dist
```

In the case of the geographical coordinates longitude and latitude, the geospatial coordinates can be added to the nodes, specifying the WGS84 coordinate system. In this example, two students were added to the graph and the knows relationship was established between them. In this case, the two nodes in the graph can be seen following a tabular representation of the data, rather than a graph-based one (Figure 3-36).

```
CREATE (peter:Student {name: 'Peter Ralph',
                       longitude:-8.53, latitude:41.43, crs:'wgs84'})
RETURN peter
MATCH (peter:Student {name: 'Peter Ralph'})
CREATE (peter) - [k:knows] - > (bruce:Student {name:'Bruce Jones',
longitude:-8.79, latitude:41.53, crs:'wgs84'})
RETURN peter, k, bruce
```

peter	k	bruce

{
　"name": "Peter Ralph",
　"crs": "wgs84",
　"latitude": 41.43,
　"longitude": -8.53
}
	{
	}
		{
		　"name": "Bruce Jones",
		　"crs": "wgs84",
		　"latitude": 41.53,
		　"longitude": -8.79
		}

Figure 3-36. Tabular representation of two nodes in Neo4j.

To calculate the travel distance between two students, the following code can be used; it establishes the current location of one of the students as the origin point, and the other as the destination point. The `totalDistance` is calculated and the result is returned in meters. In this case, the students stand approximately `24 kms` apart from each other (Figure 3-37).

```
MATCH (s1:Student) -[:knows]-> (s2:Student)
WITH point({longitude:s1.longitude, latitude:s1.latitude}) AS origin,
     point({longitude:s2.longitude, latitude:s2.latitude}) AS destination
RETURN round(distance(origin, destination)) AS totalDistance
```

totalDistance
24374

Figure 3-37. Calculating the spatial distance in Neo4j.

Other spatial functions can be applied over geospatial data. For more detailed examples, please refer to the Neo4j documentation available at (Neo4j, 2018).

3.3. NEWSQL DATABASES AND TRANSLYTICAL DATABASES

The previous examples of NoSQL databases (see section 3.2) show that this type of repositories is mostly schema-free, may not comply with the ACID properties of a relational database, and may not have a SQL-based interface for querying data. Nevertheless, NoSQL databases have scalability and performance for dealing with the volume, variety, or velocity of the data available in some application domains, although they may eventually lack consistent data, transactions, and joins, which may need to be implemented by developers.

By integrating some of the most representative characteristics of relational and NoSQL databases, the scalable SQL or NewSQL paradigm adds performance and scalability to databases that are based on a relational paradigm, and comply with ACID properties and pre-defined data schemas. Existing database systems such as MySQL Cluster have benefited from these performance improvements, while several new database systems are also emerging, like VoltDB or Clustrix (Cattell, 2011).

In 2011, the 451 Group (Aslett, 2011, p.1) defined NewSQL as databases *"that deliver the scalability and flexibility promised by NoSQL while retaining the support for SQL queries and/or ACID, or to improve performance for appropriate workloads"*. NewSQL databases benefit from the value of SQL and ACID properties, and try to avoid the performance degradation that may appear when transactional operations need to be performed over data spread throughout different nodes in a system (Cattell, 2011). In a distributed system, it is a major challenge to scale databases while maintaining the ACID properties.

The Basically Available, Soft state, and Eventual consistency (BASE) properties of NoSQL databases pose some difficulties for certain application domains, particularly those that rely on data consistency. In these cases, NewSQL systems can be considered. The main features of NewSQL databases, according to Pavlo and Aslett (2016), can be summarized based on the following criteria:

- Main memory storage;

- Partitioning/sharding;

- Concurrency control;

- Secondary indexes;

- Replication;

- Crash recovery.

One additional characteristic of some NewSQL databases is the ability to perform analytical queries on recently obtained data, a feature known as real-time analytics or Hybrid Transaction-Analytical Processing (HTAP) that allows to process and analyze historical datasets and recently collected data (Pavlo & Aslett, 2016).

In some contexts, HTAP databases are also known as **translytical** databases, because they support transactional to analytical operations. Transactional databases are optimized for writing operations, thus requiring data integrity and concurrency; whereas analytical databases are optimized for reading operations which usually perform aggregations on large amounts of data. A **translytical** database can support different workloads and access patterns in real-time, using a single data tier, hence maintaining transactional integrity, performance, and scalability (Yuhanna & Gualtieri, 2016).

© FCA

This type of database supports applications that need real-time processing and insights, predictive analytics, and transactional processing. The Forrester report (Yuhanna & Gualtieri, 2016) lists the following main **translytical** databases (including NewSQL databases, NoSQL databases, or even SQL databases with extended capabilities):

- DataStax Enterprise (DataStax, 2018), a platform based on Apache Cassandra (a wide column NoSQL database), for extreme-scale applications, with a masterless and shared-nothing architecture with in-memory capabilities;

- IBM DB2 (IBM, 2018), with BLU Acceleration and in-memory technology for better performance or Analytics Accelerator (IDAA) for real-time analytics on transactional data;

- MemSQL (2018), a distributed in-memory database that runs on commodity hardware or in the cloud, which also supports analytical processing with Apache Spark;

- Microsoft SQL Server 2016 (Microsoft, 2018a) with a unified database for using in-memory column storage with in-memory processing for high performance and real-time analytics. The column store indices add performance advantages in analytics and DWing queries;

- Oracle Database In-Memory (Oracle, 2018), extending the Oracle Database 12c, supporting transactional and analytical operations on the same database. Tables are simultaneously represented in memory using the traditional row format for OLTP queries and in columnar format for analytical queries. Consistency is retained between the two storage formats;

- SAP HANA (SAP, 2018), which integrates OLTP and OLAP data processing needs, delivering low latency data access for real-time analytics, reporting, and forecasting on a single in-memory copy of the data;

- VoltDB, considered a mature NewSQL technology, which combines streaming analytics and ACID properties for data-intensive applications, offering high-throughput and low latency. It uses horizontal partitioning of the data to scale out on commodity hardware or in the cloud, supporting data replication for high availability. It can also support semi-structured data stored as JSON data structures (VoltDB, 2018). The VoltDB Community Edition is distributed in binary and source code forms under the Affero General Public License (AGPL), having also commercial licenses with additional features (VoltDB, 2018).

The way these NewSQL databases integrate transactional and analytical needs is different from a traditional BI system, where two separate databases are usually implemented, one for OLTP (with the newly generated data) and another for OLAP (historical data that is periodically refreshed). With the advent of Big Data systems, another approach was set by the Lambda Architecture (see subsection 2.4.2), where a separate batch processing system is used to compute historical data, and a stream processing system deals with the newly generated data. Currently, thanks to the HTAP approach, database systems can support high throughput and low latency demands for OLTP workloads and perform complex long-running OLAP queries on both transactional and historical data. For the NewSQL databases that were previously mentioned, SAP HANA and MemSQL classify themselves as HTAP systems. SAP HANA uses multiple execution engines, one for row-oriented data and transactional processing, and another one for column-oriented data and analytical queries. MemSQL also uses two different storage mechanisms, rows and columns, which are integrated in a single execution engine (Pavlo & Aslett, 2016).

4 OLAP-ORIENTED DATABASES FOR BIG DATA ENVIRONMENTS

INTRODUCTION

Data systems usually assume two main roles regarding data workloads: a transactional one, described in Chapter 3, and a more analytical one. In the latter, the focus is not on low latency processing for OLTP workloads but rather on performing complex and long-running OLAP queries on historical data. OLAP is a regular analytical task associated with DWs, which relies mainly on multidimensional structures that are capable of processing vast amounts of data. These structures provide interactive data processing capabilities that help decision-makers when carrying out data analytics or other exploratory tasks.

This chapter focuses on data systems that support OLAP-oriented workloads and describes Hive as the *de facto* SQL-on-Hadoop engine for Big Data environments. Hive is suited for implementing BDWs, allowing the storage, processing, and analysis of vast amounts of data. Due to its relevance in the implementation of BDWs, this chapter also addresses how dimensional models may be transformed into tabular models which can be implemented in Hive (for example, when a denormalized data model is preferred over a star schema that requires joins between tables). Finally, this chapter shows how emerging technologies such as Druid can support real-time analytics on large datasets.

4.1. HIVE: THE *DE FACTO* SQL-ON-HADOOP ENGINE

Hive, a DWing software project for Big Data environments, facilitates querying and managing large datasets stored in a distributed storage system (Hive, 2018a; Thusoo et al., 2010). Having the data in a distributed file system, Hive adds structure to the data and allows querying it through HiveQL, a SQL-like language (Capriolo, Wampler, & Rutherglen, 2012). There are three main elements used in Hive to organize the available data: **tables, partitions,** and **buckets**.

Tables are common structures with columns and rows that are stored in a HDFS directory. **Partitions** are defined on the tables to slice the data horizontally and speed up query processing. Partitions are stored in a sub-directory within a table's directory. It is thus possible to organize the data hierarchically, associating various partitions to a table. Partitions are used to prune the data during a specific query; they have a strong impact on the time required for a query to be processed. **Buckets** can be defined and associated to tables or partitions; they are stored in a file within the partition or table's directory, depending on whether the table is partitioned or not. Buckets are used to cluster large datasets and optimize query performance. When a table is created, the user can specify both the number of buckets and the column that will be used for bucketing the data. Once again, this technique is used to prune the data in case the user runs a query on a sample of the available data (Capriolo et al., 2012).

As an example, one can assume that the table `Employees` is created in Hive and includes five attributes, as illustrated in Figure 4-1. In this case, the available data is stored in a single file in HDFS, under the `myData` database directory, assuming this is the working database.

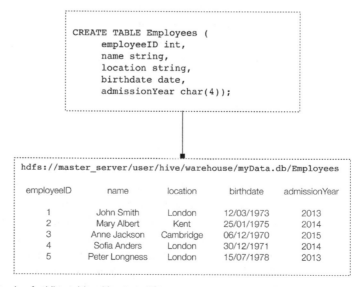

Figure 4-1. Example of a Hive table without partitions.

If the attribute `admissionYear` is used as a partition, the directory of the table will contain as many subdirectories as the number of years of admission listed in the dataset, splitting the data into several files (Figure 4-2). If a `where` condition is used while querying the data, to filter for instance a specific year, only the subdirectory associated to that specific condition is searched.

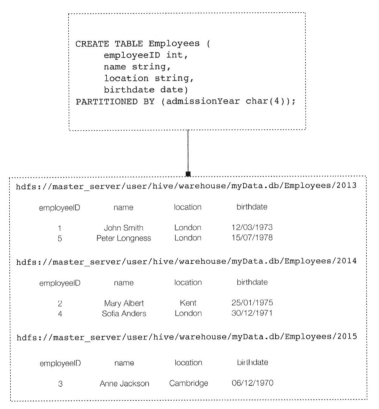

```
CREATE TABLE Employees (
     employeeID int,
     name string,
     location string,
     birthdate date)
PARTITIONED BY (admissionYear char(4));
```

hdfs://master_server/user/hive/warehouse/myData.db/Employees/2013

employeeID	name	location	birthdate
1	John Smith	London	12/03/1973
5	Peter Longness	London	15/07/1978

hdfs://master_server/user/hive/warehouse/myData.db/Employees/2014

employeeID	name	location	birthdate
2	Mary Albert	Kent	25/01/1975
4	Sofia Anders	London	30/12/1971

hdfs://master_server/user/hive/warehouse/myData.db/Employees/2015

employeeID	name	location	birthdate
3	Anne Jackson	Cambridge	06/12/1970

Figure 4-2. Example of a Hive table with partitions.

Partitions can have subpartitions or buckets. Subpartitions provide a hierarchical division of the storage space, whereas buckets organize the available data in files according to the predefined number of buckets. Taking as an example the `location` and the `admissionYear` of the employee, Figure 4-3 shows how data is organized according to the available values for the defined partition and subpartition.

Having shown a few short examples on how data can be organized in Hive, it is important to mention that the performance of a Hive data model depends on the partitions and buckets. The definition of partitions and buckets is constrained by the available data, for instance, the number of different employees, locations, or years. Data should not be over partitioned. If partitions are relatively small, compared to the data volume, searching many

© FCA

directories with small files typically slows down the processing time of Hadoop. Partitions should be of similar size to prevent, for example, that a single long-running operation is carried out in an excessively large partition. Returning for a moment to the location attribute mentioned above, let us suppose that more than 80% of the employees are from two or three different cities. Partitioning the data using this attribute would not be adequate because the majority of the data would be stored in only two or three directories. In this case, bucketing may not be an alternative as well, as many more rows would be clustered in the same bucket. However, bucketing can be useful for attributes with high cardinality and a well-balanced distribution within the attributes range. In that case, joining two Hive tables using the location attribute, for instance, allows Hive to join bucket by bucket; this operation performs even better if the buckets are sorted according to the values of the location attribute.

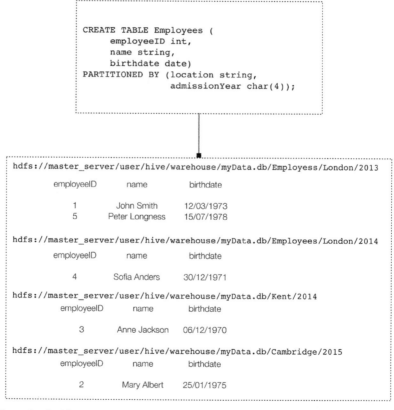

Figure 4-3. Example of a Hive table with partitions and subpartitions.

In general, partitioning helps to reduce the amount of data that must be searched, when the user applies a condition in a `where` clause; whereas bucketing helps to organize the data inside each table or partition using multiple files. The same subset of data is always written in the same bucket, which facilitates the task of joining columns. Usually, bucketing works well when the attribute has high cardinality, and partitioning when the cardinality is not too high and the available rows are well distributed by the different attribute's values. Partitioning can be applied to multiple attributes, whereas bucketing can only be applied to a single one. Later in this chapter (see subsection 4.1.2), more detailed examples of how to create partitions, subpartitions, and buckets are provided. How these design patterns impact performance and how the different data models behave are analyzed in more detail in Chapter 8.

Tables can be created using the Hive CLI, where the HiveQL supports a SQL-based language to perform Data Definition Language (DDL) and Data Manipulation Language (DML) operations, but they can also be created through any application that supports data preparation and transformation tasks.

Tables in Hive can be internal or external, depending on their goal. **Internal Hive tables** are managed by Hive and their access is restricted to a specific user account, database, or even data schema. When an internal table is deleted, both the associated metadata and the data are deleted from Hive and from HDFS. These internal tables are recommended for storing data temporarily or when the management of the tables' life cycle and their corresponding data is left entirely to Hive. **External Hive tables** can be accessed by users who also have access to HDFS, so security needs to be managed at the HDFS file/folder level. When an external table is deleted, Hive only removes its metadata and not the files themselves, which remain available in HDFS. External tables are normally used when data needs to be managed outside Hive, for instance in ETL/ELT processes. External tables are also useful when there is the need to maintain the history of the files used to load or refresh a DW, given that temporary tables are deleted after the updating process, but all the corresponding files remain within HDFS. The following subsections provide more detailed descriptions and examples of Hive tables.

In the following examples, the Star Schema Benchmark (SSB) data model is used (O'Neil, O'Neil, & Chen, 2009). This model includes a central fact table for the line items ("`lineorder`") associated with the orders, which are related to four dimension tables, "`supplier`", "`customer`", "`part`", and "`date`" (Figure 4-4). The SSB is often used for benchmarking purposes, as shown later in this book, because it can be tested using different SFs to generate datasets of different sizes.

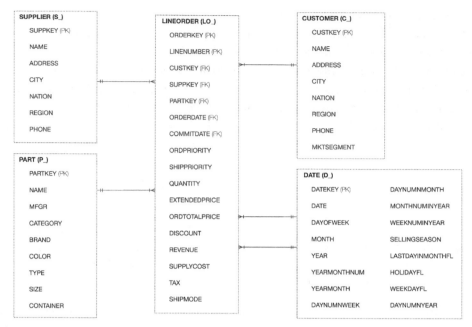

Figure 4-4. SSB data model. Adapted from O'Neil et al. (2009).

4.1.1. Data Storage Formats

The SSB data model with a SF of 10 feeds a `customer` table with 1,500,000 rows. This table can be created using the following DDL code, which specifies the table's name, attributes, row delimiter, format (both input and output), and location. In this case, the table is stored in text format – the default value – and created as external, as it is later used in this subsection to feed other Hive tables with different file formats. The table is stored inside a specific location in HDFS, where the `hivessb` database was made available in the `/user/lid4/book` folder for the `lid4` user.

```
CREATE EXTERNAL TABLE `customer`(
  `custkey` string,
  `name` string,
  `address` string,
  `city` string,
  `nation` string,
  `region` string,
  `phone` string,
  `mktsegment` string)
ROW FORMAT DELIMITED
  FIELDS TERMINATED BY '\;'
STORED AS INPUTFORMAT
  'org.apache.hadoop.mapred.TextInputFormat'
```

```
OUTPUTFORMAT
   'org.apache.hadoop.hive.ql.io.HiveIgnoreKeyTextOutputFormat'
LOCATION

'hdfs://yourservername:8020/user/lid4/book/hivessb.db/customer'
```

After creating and loading the data, the user can use the DML to make queries. Figure 4-5 shows a query selecting the customers of a specific nation. In this case, the Ambari view for Hive was used both for posting the query and for receiving the results.

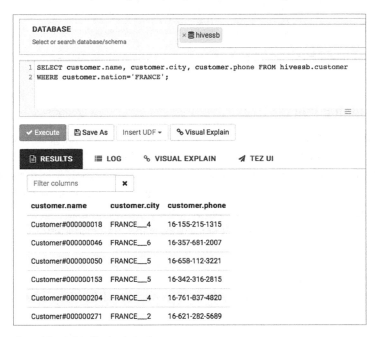

Figure 4-5. Querying a Hive table (Ambari view).

File formats supported by Hive have evolved over time to enhance the analytical capabilities of this system, given that file formats have a strong influence over the size of the tables and the performance of the queries carried on them. Some example of file formats are text file, sequence file, Record Columnar File (RCFile), ORC file, Avro file, and Parquet; the custom INPUTFORMAT and OUTPUTFORMAT options allow users to define how to read complex files and convert them to other formats.

4.1.1.1. Text File

The text file format was shown in the previous example. In this case, the data contained in the Hive table is stored as txt in HDFS. This is the default storage format in Hive and it can support any text file with a delimiter set by the user.

Following the previous example, Figure 4-6 shows the location of the file in HDFS, alongside a small extract of the table, where some rows, their attributes, and the defined delimiter ";" can be seen.

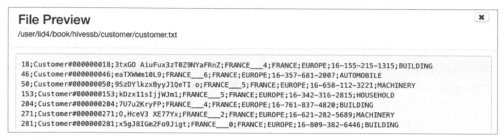

File Preview   ✖
/user/lid4/book/hivessb/customer/customer.txt

```
18;Customer#000000018;3txGO AiuFux3zT0Z9NYaFRnZ;FRANCE___4;FRANCE;EUROPE;16-155-215-1315;BUILDING
46;Customer#000000046;eaTXWWm10L9;FRANCE___6;FRANCE;EUROPE;16-357-681-2007;AUTOMOBILE
50;Customer#000000050;9SzDYlkzxByyJ1QeTI o;FRANCE___5;FRANCE;EUROPE;16-658-112-3221;MACHINERY
153;Customer#000000153;kDzx11sIjjWJm1;FRANCE___5;FRANCE;EUROPE;16-342-316-2815;HOUSEHOLD
204;Customer#000000204;7U7u2KryFP;FRANCE___4;FRANCE;EUROPE;16-761-837-4820;BUILDING
271;Customer#000000271;0,HceV3 XE77Yx;FRANCE___2;FRANCE;EUROPE;16-621-282-5689;MACHINERY
281;Customer#000000281;x5gJ8IGm2Fo9Jigt;FRANCE___0;FRANCE;EUROPE;16-809-382-6446;BUILDING
```

Figure 4-6. Overview of the text file of a Hive table.

4.1.1.2. Sequence File

Sequence files are flat files with binary key-value pairs, intended for grouping two or more smaller files and transforming them into a sequence file. If a file's size is smaller than a typical block size in Hadoop (usually 64MB in Apache Hadoop), it is recommended to merge it with other similarly small files to form a larger one, because this will positively impact Hadoop's performance. Using sequence files when creating Hive tables is a good method for generating such larger files since they serve as containers for storing smaller files (White, 2015).

Sequence files include a header followed by one or more rows. The header contains specific information about the file, such as the names of the key and value classes, compression details, user-defined metadata, and the sync marker. The sync marker is a randomly generated value for each file, which is spread among rows (records) in the sequence file. Since they are designed to add an overall storage overhead of no more than 1%, they do not necessarily exist between every pair of records (Figure 4-7). Sync points allow files to be split and sent to different maps in MapReduce jobs; they can also be used to resynchronize a record after searching a position in the data stream, given that sync points are aligned with records boundaries (White, 2015).

When storing data in a sequence file, the internal format of the records depends on factors such as whether compression is enabled or not. In case it is, record compression or block compression can be applied. No compression is the default value. In this case, each record in the sequence file includes the record length, the key length, the key itself, and the value. Lengths are usually written as 4-byte integers. When compression is enabled, the file format is similar to the above, but the value is compressed using the coder/decoder defined in the header (Figure 4-7) (White, 2015).

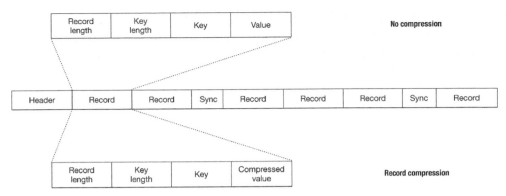

Figure 4-7. Sequence files with no compression or with record compression. Adapted from White (2015).

In the case of block compression, multiple records are stored together based on their similarities, thus providing a more compact way of organizing the data. A block compacts records until they reach a given size in bytes (defined in the block size properties), and a sync mark is written at the beginning of each block. The format of a block includes the number of records contained within itself along with four compressed fields: the key lengths, the keys, the value lengths, and the values (Figure 4-8) (White, 2015).

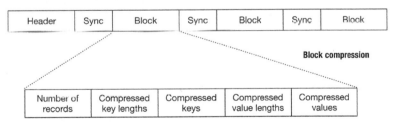

Figure 4-8. Sequence files with block compression. Adapted from White (2015).

The following example shows how the sequence file can be created and the way it is organized. In this case, the customer table that was previously created using the text file format is now used to feed a table using the sequence file format. This table is loaded using the INSERT OVERWRITE TABLE instruction, which inserts the data selected from the text file table in the recently created table.

```
CREATE TABLE hivessb.customerseq
    LIKE hivessb.customer
    STORED AS SEQUENCEFILE;


INSERT OVERWRITE TABLE hivessb.customerseq
    SELECT * FROM hivessb.customer;
```

Nine different files are created, spreading the data into several files that complement each other. The system shell is used to invoke specific HDFS commands such as `dfs` for running a file system command in HDFS and `-ls` for listing the contents of the `customerseq` folder associated with the table.

```
$ hdfs dfs -ls /user/lid4/book/hivessb.db/customerseq
Found 9 items
-rwxr-xr-x   2 lid4 18936908 2018-03-14 19:20 /user/lid4/book/hivessb.db/customerseq/000000_0
-rwxr-xr-x   2 lid4 18959090 2018-03-14 19:20 /user/lid4/book/hivessb.db/customerseq/000001_0
-rwxr-xr-x   2 lid4 18980136 2018-03-14 19:20 /user/lid4/book/hivessb.db/customerseq/000002_0
-rwxr-xr-x   2 lid4 18938791 2018-03-14 19:20 /user/lid4/book/hivessb.db/customerseq/000003_0
-rwxr-xr-x   2 lid4 18929553 2018-03-14 19:20 /user/lid4/book/hivessb.db/customerseq/000004_0
-rwxr-xr-x   2 lid4 18916126 2018-03-14 19:20 /user/lid4/book/hivessb.db/customerseq/000005_0
-rwxr-xr-x   2 lid4 18896068 2018-03-14 19:20 /user/lid4/book/hivessb.db/customerseq/000006_0
-rwxr-xr-x   2 lid4 18966051 2018-03-14 19:20 /user/lid4/book/hivessb.db/customerseq/000007_0
-rwxr-xr-x   2 lid4 20029489 2018-03-14 19:20 /user/lid4/book/hivessb.db/customerseq/000008_0
```

The `-cat` command may be used to inspect files and their contents can be listed using `-ls`. In this case, file `000000_0` is inspected; the first 3 bytes (`SEQ`) tell us this is a sequence file that uses the key class `org.apache.hadoop.io.BytesWritable` and the value class `org.apache.hadoop.io.Text`. In this case, no compression was defined. Otherwise, the type of compression would also be listed.

```
$ hdfs dfs -cat /user/lid4/book/hivessb.db/customerseq/000000_0
SEQ"org.apache.hadoop.io.BytesWritableorg.apache.hadoop.io.Text??s?

?FA?sD?????fa18;Customer#000000018;3txGO AiuFux3zTOZ9NYaFRnZ;FRANCE___4;FRANCE;EUROPE;16-155-215-
1315;BUILDINGZU46;Customer#000000046;eaTXWWm1OL9;FRANCE___6;FRANCE;EUROPE;16-357-681-
2007;AUTOMOBILEb]50;Customer#000000050;9SzDYlkzxByyJlQeTI o;FRANCE___5;FRANCE;EUROPE;16-658-112-
3221;MACHINERY]X153;Customer#000000153;kDzx11sIjjWJm1;FRANCE___5;FRANCE;EUROPE;16-342-316-
2815;HOUSEHOLDXS204;Customer#000000204;7U7u2KryFP;FRANCE___4;FRANCE;EUROPE;16-761-837-
4820;BUILDING]X271;Customer#000000271;O,HceV3 XE77Yx;FRANCE___2;FRANCE;EUROPE;16-621-282-
5689;MACHINERY^Y281;Customer#000000281;x5gJ8IGm2Fo9Jigt;FRANCE___0;FRANCE;EUROPE;16-809-382-
6446;BUILDINGid284;Customer#000000284;2ZgAkaBgb6aigORfIfUd3kHbP;FRANCE___2;FRANCE;EUROPE;16-161-235-
2690;AUTOMOBILEZU312;Customer#000000312;cH6XucXV0V;FRANCE___1;FRANCE;EUROPE;16-316-482-
2555;AUTOMOBILE\W …
```

If we take a longer extract of the listed data, the sync marks between the records, which allow the data to be distributed (as was already explained), become visible.

```
SEQ"org.apache.hadoop.io.BytesWritableorg.apache.hadoop.io.Text??s?

?FA?sD?????fa18;Customer#000000018;3txGO AiuFux3zTOZ9NYaFRnZ;FRANCE___4;FRANCE;EUROPE;16-155-215-
1315;BUILDINGZU46;Customer#000000046;eaTXWWm1OL9;FRANCE___6;FRANCE;EUROPE;16-357-681-
2007;AUTOMOBILEb]50;Customer#000000050;9SzDYlkzxByyJlQeTI o;FRANCE___5;FRANCE;EUROPE;16-658-112-
3221;MACHINERY]X153;Customer#000000153;kDzx11sIjjWJm1;FRANCE___5;FRANCE;EUROPE;16-342-316-
2815;HOUSEHOLDXS204;Customer#000000204;7U7u2KryFP;FRANCE___4;FRANCE;EUROPE;16-761-837-
4820;BUILDING]X271;Customer#000000271;O,HceV3 XE77Yx;FRANCE___2;FRANCE;EUROPE;16-621-282-
5689;MACHINERY^Y281;Customer#000000281;x5gJ8IGm2Fo9Jigt;FRANCE___0;FRANCE;EUROPE;16-809-382-
6446;BUILDINGid284;Customer#000000284;2ZgAkaBgb6aigORfIfUd3kHbP;FRANCE___2;FRANCE;EUROPE;16-161-235-
2690;AUTOMOBILEZU312;Customer#000000312;cH6XucXV0V;FRANCE___1;FRANCE;EUROPE;16-316-482-
```

```
2555;AUTOMOBILE\W319;Customer#000000319; UQ5mF3sdoZT2;FRANCE___3;FRANCE;EUROPE;16-734-928-
1642;FURNITUREid402;Customer#000000402;8Cw4p1m1gKYVUgomkAq,es1Zt;FRANCE___1;FRANCE;EUROPE;16-950-729-
1638;AUTOMOBILEc^413;Customer#000000413;,4Jm5N0ruhJCB7cBR6Kw;FRANCE___5;FRANCE;EUROPE;16-158-285-
7336;FURNITURE_Z431;Customer#000000431;RNfSXbUJkgUlBBPn;FRANCE___3;FRANCE;EUROPE;16-326-904-
6643;HOUSEHOLDa\452;Customer#000000452;,TI7FdTc gCXUMi09qD;FRANCE___3;FRANCE;EUROPE;16-335-974-
9174;BUILDING^Y455;Customer#000000455;sssuscPJ,ZYQ8viO;FRANCE___5;FRANCE;EUR OPE;16-863-225-
9454;BUILDINGhc459;Customer#000000459;CkGH34iK 9vAHXeY7 wAQIzJa;FRANCE___0;FRANCE;EUROPE;16-927-662-
8584;MACHINERYhc516;Customer#000000516;EJwOQMTQnFwvd8r Y7f9i5POy;FRANCE___1;FRANCE;EUROPE;16-947-309-
2690;MACHINERYhc539;Customer#000000539;FoGcDu9llpFiB LELF3rdjaiw;FRANCE___6;FRANCE;EUROPE;16-166-785-
8571;HOUSEHOLDid587;Customer#000000587;J2UwoJEQzAOTtuBrxGVag9iWS;FRANCE___4;FRANCE;EUROPE;16-585-233-
5906;AUTOMOBILE??????s?

?FA?sD?????id667;Customer#000000667;oQqeEC,OD9XC1JXyOsHqcpv0f;FRANCE___5;FRANCE;EUROPE;16-917-453-
2490;AUTOMOBILEhc683;Customer#000000683;G0, q8c6vBykpiLvcuSJLYvqE;FRANCE___3;FRANCE;EUROPE;16-566-251-
5446;MACHINERY`[686;Customer#000000686;1j C80VWHe ITCVCV;FRANCE___1;FRANCE;EUROPE;16-682-293-
3599;HOUSEHOLDid703;Customer#000000703;ge1GEYt4ewGUiSeqBA4rNB5Jh;FRANCE___0;FRANCE;EUROPE;16-741-513-
6919;AUTOMOBILEa\712;Customer\000000712; 8w2pIiA4wWAhtjAdXR;FRANCE___6;FRANCE;EUROPE;16-843-486-
5087;BUILDINGhc833;Customer#000000833;t3qDCo,Yh MZcJFV6PiheY,MU;FRANCE___6;FRANCE;EUROPE;16-624-307-
4875;FURNITURE\W873;Customer#000000873;XFnr9C2bANXL;FRANCE___4;FRANCE;EUROPE;16-375-385-
5712;AUTOMOBILEb]897;Customer#000000897;nW1X1H19uWycuBEu3F3;FRANCE___0;FRANCE;EUROPE;16-988-776-
4568;MACHINERYZU906;Customer#000000906;lUavkms1A5z;FRANCE___3;FRANCE;EUROPE;16-594-569-
6627;HOUSEHOLDid1021;Customer#000001021;m h2wQbujQnQOrcf109reW0 o;FRANCE___6;FRANCE;EUROPE;16-469-554-
5196;MACHINERYid1097;Customer#000001097;a wMc01QutcHs6cRomoMCGjvM;FRANCE___6;FRANCE;EUROPE;16-604-758-
5574;MACHINERY]X1111;Customer#000001111;gavpg6eW5lEML;FRANCE___1;FRANCE;EUROPE;16-824-312-
3537;MACHINERY_Z1182;Customer#000001182;pLrF7F1,uoyGaU;FRANCE___1;FRANCE;EUROPE;16-229-473-
7194;AUTOMOBILEhc1238;Customer#000001238;HGCJI27,RIIQcS20,DcJbMQuU;FRANCE___8;FRANCE;EUROPE;16-302-
171-7578;BUILDINGgb1379;Customer#000001379;rqYSBCMywMKnfcp2DwotVqI;FRANCE___6;FRANCE;EUROPE;16-695-
982-9623;MACHINERYfa1402;Customer#000001402;F7 m0JwiCABmbJLPQpCJ2;FRANCE___4;FRANCE;EUROPE;16-713-144-
2780;AUTOMOBILEfa1477;Customer#000001477;nUT6kGEr7tmgpJaPgfFtXY;FRANCE___3;FRANCE;EUROPE;16-407-756-
8079;MACHINERYfa1569;Customer#000001569;4vO9w7ixKJ 5od18LqLvr,;FRANCE___5;FRANCE;EUROPE;16-108-793-
2841;HOUSEHOLD_Z1571;Customer#000001571;akbtXy3o6igP3n8C;FRANCE___5;FRANCE;EUROPE;16-661-716-
6605;BUILDING??????s?

?FA?sD?????id1604;Customer#000001604;DXn5Lr8KjjMebZznHhSsX3n7T;FRANCE___1;FRANCE;EUROPE;16-960-140-
9357;HOUSEHOLDhc1608;Customer#000001608;jGjdmzMbF05pXU5STryOYpL9o;FRANCE___4;FRANCE;EUROPE;16-897-134-
9884;BUILDINGid1669;Customer#000001669;i38,,EDjrqVpk1UKXs19cCdAw;FRANCE___8;FRANCE;EUROPE;16-172-628-
3560;HOUSEHOLDlg281258;Customer#000281258;O6YY,1W3VB9gZrp5pYT0hG02J;FRANCE___7;FRANCE;EUROPE;16-956-
985-3604;AUTOMOBILElg281291;Customer#000281291;sy2xVPrxxVtEejUXLCGDrExnn;FRANCE___5;FRAN CE;EUROPE;16-
335-832-2943;AUTOMOBILEc^281299;Customer#000281299;Z mkvsf3QptL7BVi1;FRANCE___4;FRANCE;EUROPE;16-753-
579-8545;FURNITURE]X281346;Customer#000281346;KlYhkEcj3T;FRANCE___7;FRANCE;EUROPE;16-420-442-
7414;AUTOMOBILEid281384;Customer#000281384;rv39QThJOsX0nZiUHOdRnlf;FRANCE___0;FRANCE;EUROPE;16-463-
229-2525;FURNITURElg281416;Customer#000281416;d,rYZvEL0TxWtCfzLP 5u,Rq9;FRANCE___0;FRANCE;EUROPE;16-
659-402-2709;AUTOMOBILEkf281441;Customer#000281441;aBLvsY,LioJqE1DzM z8AHdeR;FRANCE___1;FRANCE;EUROPE;16-
947-800-1862;FURNITURE_Z281484;Customer#000281484;qjTGdDEJeZt0O;FRANCE___4;FRANCE;EUROPE;16-396-270-
1594;MACHINERYkf281497;Customer#000281497;uRGeT3IFtorqoUmQg,URmTpkR;FRANCE___4;FRANCE;EUROPE;16-119-
692-8659;HOUSEHOLDhc281540;Customer#000281540;h7mIVxRdaqb3 WLOuXSEpN;FRANCE___4;FRANCE;EUROPE;16-701-
716-3981;HOUSEHOLDje281560;Customer#000281560;fhh2tqN35,Z7FA9SSep0J6Ggy;FRANCE___1;FRANCE;EUROPE;16-
211-553-5607;BUILDINGfa281577;Customer#000281577;IZYOzjuqp Fj7SdsgSv4;FRANCE___8;FRANCE;EUROPE;16-638-
846-8126;HOUSEHOLDa\281605;Customer#000281605;6CtLD,D4Le3gzi;FRANCE___4;FRANCE;EUROPE;16-217-711-
5552;AUTOMOBILEkf281617;Customer#000281617;LvVz,nQCmLD7GR4KsrQMiqkW5;FRANCE___0;FRANCE;EUROPE;16-747-
114-9999;MACHINERY^Y281618;Customer#000281618;QcrXat54b2Q3;FRANCE___8;FRANCE;EUROPE;16-571-692-
3521;FURNITUREkf281642;Customer#000281642;ZWeRT78BcFzUqLhaPVC2cLk61;FRANCE___5;FRANCE;EUROPE;16-641-
903-4096;FURNITURE??????s?
```

If the `dfs` command is switched from `-cat` to `-text`, the contents of the source file are
shown in plain text format, however, both the line separators and the sync marks are
omitted.

```
$ hdfs dfs -text /user/lid4/book/hivessb.db/customerseq/000000_0
18;Customer#000000018;3txGO AiuFux3zT0Z9NYaFRnZ;FRANCE___4;FRANCE;EUROPE;16-155-215-1315;BUILDING
46;Customer#000000046;eaTXWWm10L9;FRANCE___6;FRANCE;EUROPE;16-357-681-2007;AUTOMOBILE
50;Customer#000000050;9SzDYlkzxByyJ1QeTI o;FRANCE___5;FRANCE;EUROPE;16-658-112-3221;MACHINERY
153;Customer#000000153;kDzx11sIjjWJm1;FRANCE___5;FRANCE;EUROPE;16-342-316-2815;HOUSEHOLD
204;Customer#000000204;7U7u2KryFP;FRANCE___4;FRANCE;EUROPE;16-761-837-4820;BUILDING
271;Customer#000000271;O,HceV3 XE77Yx;FRANCE___2;FRANCE;EUROPE;16-621-282-5689;MACHINERY
281;Customer#000000281;x5gJ8IGm2Fo9Jigt;FRANCE___0;FRANCE;EUROPE;16-809-382-6446;BUILDING
284;Customer#000000284;2ZgAkaBgb6aigORfIfUd3kHbP;FRANCE___2;FRANCE;EUROPE;16-161-235-2690;AUTOMOBILE
312;Customer#000000312;cH6XucXV0V;FRANCE___1;FRANCE;EUROPE;16-316-482-2555;AUTOMOBILE
319;Customer#000000319; UQ5mF3sdoZT2;FRANCE___3;FRANCE;EUROPE;16-734-928-1642;FURNITURE
402;Customer#000000402;8Cw4p1m1gKYVUgomkAq,es1Zt;FRANCE___1;FRANCE;EUROPE;16-950-729-1638;AUTOMOBILE
413;Customer#000000413;,4Jm5N0ruhJCB7cBR6Kw;FRANCE___5;FRANCE;EUROPE;16-158-285-7336;FURNITURE
431;Customer#000000431;RNfSXbUJkgUlBBPn;FRANCE___3;FRANCE;EUROPE;16-326-904-6643;HOUSEHOLD
452;Customer#000000452;,TI7FdTc gCXUMi09qD;FRANCE___3;FRANCE;EUROPE;16-335-974-9174;BUILDING
455;Customer#000000455;sssuscPJ,ZYQ8viO;FRANCE___5;FRANCE;EUROPE;16-863-225-9454;BUILDING
459;Customer#000000459;CkGH34iK 9vAHXeY7 wAQIzJa;FRANCE___0;FRANCE;EUROPE;16-927-662-8584;MACHINERY
516;Customer#000000516;EJwOQMTQnFwvd8r Y7f9i5POy;FRANCE___1;FRANCE;EUROPE;16-947-309-2690;MACHINERY
539;Customer#000000539;FoGcDu9llpFiB LELF3rdjaiw;FRANCE___6;FRANCE;EUROPE;16-166-785-8571;HOUSEHOLD
...
```

RCFiles (see subsection 4.1.1.3) can be created following the same approach as for sequence files. This format is useful for implementing DWing systems based on MapReduce (Hive, 2018c).

```
CREATE TABLE `customerrcf`(
  `custkey` string,
  `name` string,
  `address` string,
  `city` string,
  `nation` string,
  `region` string,
  `phone` string,
  `mktsegment` string)
ROW FORMAT DELIMITED
  FIELDS TERMINATED BY '\;'
STORED AS INPUTFORMAT
  'org.apache.hadoop.hive.ql.io.RCFileInputFormat'
OUTPUTFORMAT
  'org.apache.hadoop.hive.ql.io.RCFileOutputFormat'
LOCATION
  'hdfs://yourservername:8020/user/lid4/book/hivessb.db/customerrcf'
INSERT OVERWRITE TABLE hivessb.customerrcf
  SELECT * FROM hivessb.customer;
```

Sequence files are row-oriented, hence the values in a row are stored contiguously in the file; whereas columnar-oriented formats split the rows so the resulting splits may be stored in a columnar way. The values of each row in the first column are stored first, followed by the values of each row in the second column, and so on, continuing this process for all available rows/columns (White, 2015).

4.1.1.3. RCFile

Hive's RCFiles were the first column-oriented file format to be used in Hadoop (White, 2015) and are implemented through binary key-value pairs. This type of file combines a horizontal and a vertical data partition, taking advantage of row-store and column-store features for fast data loading and query processing, which also ensures the storage space is used efficiently.

This storage approach first partitions rows horizontally into row splits, and then vertically into columns. The metadata of the row split is stored in the key, whereas the data of the row split is stored in the value. With horizontal partition, an RCFile ensures all data belonging to the same row is stored in the same node.

Vertical partition allows columnar data compression strategies, which help avoiding unnecessary column reads (White, 2015). Figure 4-9 shows the difference between the row-oriented layout used by a sequence file and the columnar-oriented layout used by an RCFile.

	column 1	column 2	column 3	column 4
row 1	a1	b1	c1	d1
row 2	a2	b2	c2	d2
row 3	a3	b3	c3	d3
row 4	a4	b4	c4	d4

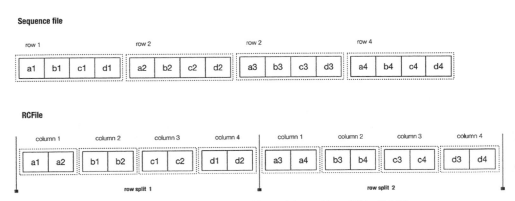

Figure 4-9. Data organization in sequence files and RCFiles. Adapted from White (2015).

Thanks to their compact structure, columnar formats allow for greater efficiency in terms of file size and query performance. Files in these formats tend to be smaller because the values in a column are stored together, thus making their encoding more efficient. The encoding of timestamps can follow a pattern that stores the first timestamp and the differences in seconds to the following ones, for example. This is normally a small difference that decreases storage needs. Since records around the same time are stored close to each other, this also decreases processing time.

In the customer table created in Hive using the RCFile format (DDL code at the end of subsection 4.1.1.2) the data is split into nine files. The same number of files was obtained when the sequence file was created.

```
$ hdfs dfs -ls /user/lid4/book/hivessb.db/customerrcf
Found 9 items
-rwxr-xr-x   2 lid4 15731756 2018-03-14 20:18 /user/lid4/book/hivessb.db/customerrcf/000000_0
-rwxr-xr-x   2 lid4 15720238 2018-03-14 20:18 /user/lid4/book/hivessb.db/customerrcf/000001_0
-rwxr-xr-x   2 lid4 15709102 2018-03-14 20:18 /user/lid4/book/hivessb.db/customerrcf/000002_0
-rwxr-xr-x   2 lid4 15730031 2018-03-14 20:18 /user/lid4/book/hivessb.db/customerrcf/000003_0
-rwxr-xr-x   2 lid4 15735497 2018-03-14 20:18 /user/lid4/book/hivessb.db/customerrcf/000004_0
-rwxr-xr-x   2 lid4 15741607 2018-03-14 20:18 /user/lid4/book/hivessb.db/customerrcf/000005_0
-rwxr-xr-x   2 lid4 15752128 2018-03-14 20:18 /user/lid4/book/hivessb.db/customerrcf/000006_0
-rwxr-xr-x   2 lid4 15715629 2018-03-14 20:18 /user/lid4/book/hivessb.db/customerrcf/000007_0
-rwxr-xr-x   2 lid4 16538443 2018-03-14 20:18 /user/lid4/book/hivessb.db/customerrcf/000008_0
```

When one of the files is inspected, it shows that the organization of the data differs from the previous formats (text and sequence files), since data values are grouped in columns to optimize storage space and increase performance when processing the data. The following example illustrates the transition of the values from the `region` column to the `phone` column.

```
$ hdfs dfs -text /user/lid4/book/hivessb.db/customerrcf/000000_0
…
EUROPEEUROPEEUROPEEUROPEEUROPEEUROPEEUROPEEUROPEEUROPEEUROPEEUROPEEUROPEEUROPEEUROPEEUROPEEUROPEEUROPEEUR
OPEEUROPEEUROPEEUROPEEUROPEEUROPEEUROPEEUROPEEUROPEEUROPEEUROPEEUROPEEUROPEEUROPEEUROPEEUROPEEUROPEEUROPE
EUROPEEUROPEEUROPEEUROPEEUROPEEUROPEEUROPEEUROPEEUROPEEUROPEEUROPEEUROPEEUROPEEUROPEEUROPEEUROPEEUROPEEUR
OPEEUROPEEUROPEEUROPEEUROPEEUROPEEUROPEEUROPEEUROPEEUROPEEUROPEEUROPEEUROPEEUROPEEUROPEEUROPEEUROPEEUROPE
EUROPEEUROPEEUROPEEUROPEEUROPEEUROPEEUROPEEUROPEEUROPEEUROPEEUROPEEUROPEEUROPEEUROPEEUROPEEUROPEEUROPEEUR
OPEEUROPEEUROPEEUROPEEUROPEEUROPEEUROPEEUROPEEUROPEEUROPEEUROPEEUROPEEUROPEEUROPEEUROPEEUROPEEUROPEEUROPE
EUROPEEUROPEEUROPEEUROPEEUROPEEUROPEEUROPEEUROPEEUROPEEUROPEEUROPEEUROPEEUROPEEUROPEEUROPEEUROPEEUROPEEUR
OPEEUROPEEUROPEEUROPEEUROPEEUROPEEUROPEEUROPE 33-952-259-179133-765-257-727833-571-530-437933-877-472-
109233-142-677-600633-375-743-912833-474-727-669433-617-160-576633-419-459-882433-373-790-792233-719-447-
182733-504-255-853333-605-771-468333-309-434 157533-365-620-981233-788-834-754633-388-899-813533-349-354-
728233-998-878-378333-942-761-458133-819-124-949033-829-912-447633-908-345-127233-565-201-340133-801-768-
366033-316-941-562833-982-549-945433-228-977-241233-435-476-494933-829-932-501833-707-219-780433-240-166-
652233-352-130-254333-123-586-746833-398-157-389433-406-209-818333-399-366-995733-448-107-140733-532-932-
544233-966-203-110633-947-537-681433-419-952-996733-904-993-140233-315-169-270933-676-713-777933-583-632-
81123 …
```

Another approach for inspecting the data uses Hive's `rcfilecat` command; in this case, the data is displayed in row format, where the values of the columns are grouped according to their corresponding row.

```
$ hive --rcfilecat /user/lid4/book/hivessb.db/customerrcf/000000_0
18  Customer#000000018 3txGO AiuFux3zTOZ9NYaFRnZ FRANCE___4 FRANCE EUROPE 16-155-215-1315 BUILDING
46  Customer#000000046 eaTXWWm1OL9 FRANCE___6 FRANCE EUROPE 16-357-681-2007 AUTOMOBILE
50  Customer#000000050 9SzDYlkzxByyJ1QeTI o  FRANCE___5 FRANCE EUROPE 16-658-112-3221 MACHINERY
153 Customer#000000153 kDzx11sIjjWJm1 FRANCE___5 FRANCE EUROPE 16-342-316-2815 HOUSEHOLD
204 Customer#000000204 7U7u2KryFP FRANCE___4 FRANCE EUROPE 16-761-837-4820 BUILDING
271 Customer#000000271 O,HceV3 XE77Yx FRANCE___2 FRANCE EUROPE 16-621-282-5689 MACHINERY
281 Customer#000000281 x5gJ8IGm2Fo9Jigt FRANCE___0 FRANCE EUROPE 16-809-382-6446 BUILDING
284 Customer#000000284 2ZgAkaBgb6aigORfIfUd3kHbP FRANCE___2 FRANCE EUROPE 16-161-235-2690 AUTOMOBILE
312 Customer#000000312 cH6XucXV0V FRANCE___1 FRANCE EUROPE 16-316-482-2555 AUTOMOBILE
319 Customer#000000319 UQ5mF3sdoZT2 FRANCE___3 FRANCE EUROPE 16-734-928-1642 FURNITURE
402 Customer#000000402 8Cw4p1m1gKYVUgomkAq,es1Zt FRANCE___1 FRANCE EUROPE 16-950-729-1638 AUTOMOBILE
413 Customer#000000413 ,4Jm5N0ruhJCB7cBR6Kw FRANCE___5 FRANCE EUROPE 16-158-285-7336 FURNITURE
431 Customer#000000431 RNfSXbUJkgUlBBPn  FRANCE___3 FRANCE EUROPE 16-326-904-6643 HOUSEHOLD
452 Customer#000000452 ,TI7FdTc gCXUMi09qD FRANCE___3 FRANCE EUROPE 16-335-974-9174 BUILDING
455 Customer#000000455 sssuscPJ,ZYQ8vi0  FRANCE___5 FRANCE EUROPE 16-863-225-9454 BUILDING
459 Customer#000000459 CkGH34iK 9vAHXeY7 wAQIzJa FRANCE___0 FRANCE EUROPE 16-927-662-8584 MACHINERY
516 Customer#000000516 EJwOQMTQnFwvd8r Y7f9i5POy FRANCE___1 FRANCE EUROPE 16-947-309-2690 MACHINERY
539 Customer#000000539 FoGcDu911pFiB LELF3rdjaiw FRANCE___6 FRANCE EUROPE 16-166-785-8571 HOUSEHOLD
...
```

4.1.1.4. ORC File

The ORC file format is highly efficient for storing data in Hive, as it significantly improves performance when reading and writing data. ORC files are organized in groups of row data called "stripes", which include auxiliary information in each file's footer. At the end of each file, a postscript includes information about the compression parameters and the size of the compressed footer. The default size for each stripe is 250MB, however, larger file sizes will enable larger and more efficient reads from HDFS. Each stripe includes index data, row data, and a stripe footer with a directory of stream locations. The index data includes summaries with the min and max values and the row positions for each column. After the stripes, a file footer includes a list containing all the stripes in the file, the number of rows in each stripe, the columns' data type, and summaries of the columns using aggregation functions such as count, min, max, and sum (Figure 4-10) (Hive, 2018b).

The aggregated information about the rows stored in each stripe is useful for optimizing query performance, since the computed statistics provide valuable information about the stripes, which is needed to answer a specific query. Moreover, as can be seen in Figure 4-11, sorting the data when loading it into the stripes significantly increases the efficiency of this storage format because it reduces the number of potential stripes required to answer a given query. In the example shown in Figure 4-11, answering a query such as "SELECT * FROM customer WHERE customer id = 75" requires three stripes when data is unsorted, compared to only one stripe for the sorted version of the data.

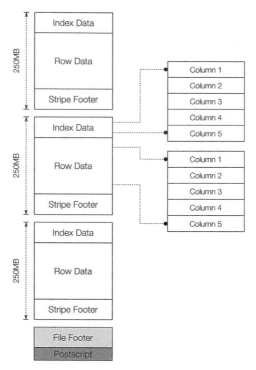

Figure 4-10. Data organization in ORC files. Adapted from Hive (2018b).

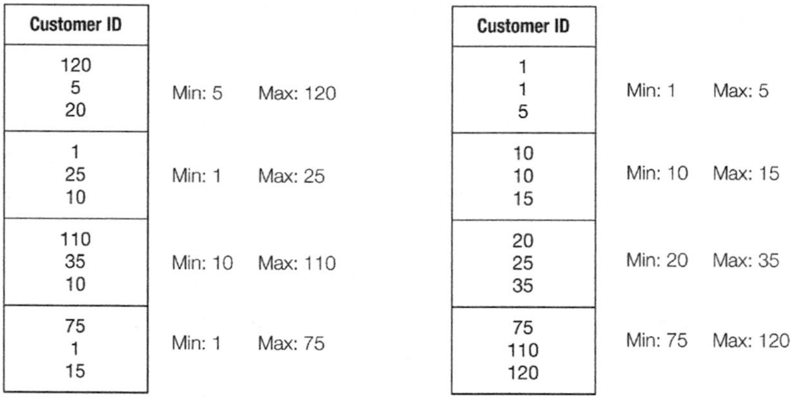

Customer ID		
120 5 20	Min: 5	Max: 120
1 25 10	Min: 1	Max: 25
110 35 10	Min: 10	Max: 110
75 1 15	Min: 1	Max: 75

Customer ID		
1 1 5	Min: 1	Max: 5
10 10 15	Min: 10	Max: 15
20 25 35	Min: 20	Max: 35
75 110 120	Min: 75	Max: 120

Figure 4-11. Unsorted and sorted data in the stripes of an ORC file.

The DDL code for creating the customer table using the ORC file format is shown below. In this case, the `txt` (text file) shown previously in Figure 4-6 is broken into the several stripes, and stored in a compressed format that substantially decreases the size of the required storage space. In the example, a 144.9MB text file converted to ORC file format distributed over nine files was reduced to approximately 44MB.

```
CREATE TABLE `customerorc`(
  `custkey` string,
  `name` string,
  `address` string,
  `city` string,
  `nation` string,
  `region` string,
  `phone` string,
  `mktsegment` string)
ROW FORMAT SERDE
  'org.apache.hadoop.hive.ql.io.orc.OrcSerde'
STORED AS INPUTFORMAT
  'org.apache.hadoop.hive.ql.io.orc.OrcInputFormat'
OUTPUTFORMAT
  'org.apache.hadoop.hive.ql.io.orc.OrcOutputFormat'
LOCATION
  'hdfs://yourservername:8020/user/lid4/book/hivessb.db/customerorc'
INSERT OVERWRITE TABLE hivessb.customerorc
  SELECT * FROM hivessb.customer;
$ hdfs dfs -ls /user/lid4/book/hivessb.db/customerorc
Found 9 items
-rwxr-xr-x   2 lid4   5083477 2018-03-12 19:59 /user/lid4/book/hivessb.db/customerorc/000000_0
-rwxr-xr-x   2 lid4   5144168 2018-03-12 19:59 /user/lid4/book/hivessb.db/customerorc/000001_0
-rwxr-xr-x   2 lid4   5192411 2018-03-12 19:59 /user/lid4/book/hivessb.db/customerorc/000002_0
-rwxr-xr-x   2 lid4   5095599 2018-03-12 19:59 /user/lid4/book/hivessb.db/customerorc/000003_0
-rwxr-xr-x   2 lid4   5065174 2018-03-12 19:59 /user/lid4/book/hivessb.db/customerorc/000004_0
-rwxr-xr-x   2 lid4   5038219 2018-03-12 19:59 /user/lid4/book/hivessb.db/customerorc/000005_0
-rwxr-xr-x   2 lid4   4986501 2018-03-12 19:59 /user/lid4/book/hivessb.db/customerorc/000006_0
-rwxr-xr-x   2 lid4   5164945 2018-03-12 19:59 /user/lid4/book/hivessb.db/customerorc/000007_0
-rwxr-xr-x   2 lid4   5545497 2018-03-12 19:59 /user/lid4/book/hivessb.db/customerorc/000008_0
```

In Hive, the `orcfiledump` command allows to inspect the contents of ORC files and their stripes. In case no specific file is indicated, for instance, when using `hive --orcfiledump /user/lid4/book/hivessb.db/customerorc`, all the files in the directory are inspected sequentially. When a specific file is inspected, the following result is obtained.

```
$ hive --orcfiledump /user/lid4/book/hivessb.db/customerorc/000000_0
Processing data file /user/lid4/book/hivessb.db/customerorc/000000_0 [length: 5083477]
Structure for /user/lid4/book/hivessb.db/customerorc/000000_0
File Version: 0.12 with HIVE_13083
18/05/28 18:39:01 INFO orc.ReaderImpl: Reading ORC rows from
/user/lid4/book/hivessb.db/customerorc/000000_0 with {include: null, offset: 0, length:
9223372036854775807}
18/05/28 18:39:01 INFO orc.RecordReaderImpl: Reader schema not provided -- using file schema
struct<_col0:string,_col1:string,_col2:string,_col3:string,_col4:string,_col5:string,_col6:string,_
col7:string>
Rows: 164665
```

© FCA

```
Compression: ZLIB
Compression size: 262144
Type:
struct<_col0:string,_col1:string,_col2:string,_col3:string,_col4:string,_col5:string,_col6:string,_
col7:string>

Stripe Statistics:
  Stripe 1:
    Column 0: count: 164665 hasNull: false
    Column 1: count: 164665 hasNull: false min: 10000 max: 999992 sum: 1027099
    Column 2: count: 164665 hasNull: false min: Customer#000000011 max: Customer#001499990 sum: 2963970
    Column 3: count: 164665 hasNull: false min:   0ii8V3kQ max: zzym,0W cT0I sum: 3480035
    Column 4: count: 164665 hasNull: false min: FRANCE___0 max: UNITED KI8 sum: 1646650
    Column 5: count: 164665 hasNull: false min: FRANCE max: UNITED KINGDOM sum: 1402446
    Column 6: count: 164665 hasNull: false min: EUROPE max: EUROPE sum: 987990
    Column 7: count: 164665 hasNull: false min: 16-100-129-9592 max: 33-999-987-8365 sum: 2469975
    Column 8: count: 164665 hasNull: false min: AUTOMOBILE max: MACHINERY sum: 1481816

File Statistics:
  Column 0: count: 164665 hasNull: false
  Column 1: count: 164665 hasNull: false min: 10000 max: 999992 sum: 1027099
  Column 2: count: 164665 hasNull: false min: Customer#000000011 max: Customer#001499990 sum: 2963970
  Column 3: count: 164665 hasNull: false min:   0ii8V3kQ max: zzym,0W cT0I sum: 3480035
  Column 4: count: 164665 hasNull: false min: FRANCE___0 max: UNITED KI8 sum: 1646650
  Column 5: count: 164665 hasNull: false min: FRANCE max: UNITED KINGDOM sum: 1402446
  Column 6: count: 164665 hasNull: false min: EUROPE max: EUROPE sum: 987990
  Column 7: count: 164665 hasNull: false min: 16-100-129-9592 max: 33-999-987-8365 sum: 2469975
  Column 8: count: 164665 hasNull: false min: AUTOMOBILE max: MACHINERY sum: 1481816

Stripes:
  Stripe: offset: 3 data: 5079200 rows: 164665 tail: 211 index: 3438
    Stream: column 0 section ROW_INDEX start: 3 length 21
    Stream: column 1 section ROW_INDEX start: 24 length 483
    Stream: column 2 section ROW_INDEX start: 507 length 484
    Stream: column 3 section ROW_INDEX start: 991 length 1056
    Stream: column 4 section ROW_INDEX start: 2047 length 214
    Stream: column 5 section ROW_INDEX start: 2261 length 248
    Stream: column 6 section ROW_INDEX start: 2509 length 162
    Stream: column 7 section ROW_INDEX start: 2671 length 548
    Stream: column 8 section ROW_INDEX start: 3219 length 222
    Stream: column 1 section DATA start: 3441 length 416253
    Stream: column 1 section LENGTH start: 419694 length 2964
    Stream: column 2 section DATA start: 422658 length 587284
    Stream: column 2 section LENGTH start: 1009942 length 1108
    Stream: column 3 section DATA start: 1011050 length 2691043
    Stream: column 3 section LENGTH start: 3702093 length 91039
    Stream: column 4 section DATA start: 3793132 length 80074
    Stream: column 4 section LENGTH start: 3873206 length 7
    Stream: column 4 section DICTIONARY_DATA start: 3873213 length 84
    Stream: column 5 section DATA start: 3873297 length 274
    Stream: column 5 section LENGTH start: 3873571 length 7
    Stream: column 5 section DICTIONARY_DATA start: 3873578 length 30
    Stream: column 6 section DATA start: 3873608 length 195
    Stream: column 6 section LENGTH start: 3873803 length 6
    Stream: column 6 section DICTIONARY_DATA start: 3873809 length 9
```

```
Stream: column 7 section DATA start: 3873818 length 1144397
Stream: column 7 section LENGTH start: 5018215 length 390
Stream: column 8 section DATA start: 5018605 length 63980
Stream: column 8 section LENGTH start: 5082585 length 8
Stream: column 8 section DICTIONARY_DATA start: 5082593 length 48
Encoding column 0: DIRECT
Encoding column 1: DIRECT_V2
Encoding column 2: DIRECT_V2
Encoding column 3: DIRECT_V2
Encoding column 4: DICTIONARY_V2[27]
Encoding column 5: DICTIONARY_V2[3]
Encoding column 6: DICTIONARY_V2[1]
Encoding column 7: DIRECT_V2
Encoding column 8: DICTIONARY_V2[5]


File length: 5083477 bytes
Padding length: 0 bytes
Padding ratio: 0%
```

4.1.1.5. Avro File

Avro files can be created and inspected following the same process described above. Avro files (Avro, 2018) use a serialization system to support nested schemas, allowing to modify a table schema across different compatible schemas, without needing to rewrite and convert existing tables. HiveQL can query different Avro files using the current table schema (Figure 4-12), since Avro files store the data and the corresponding schema. As code generation is optional in Avro, data conforming to a specific schema can be read and written, even if the code is not expecting that particular schema. This flexibility allows schemas to evolve, hence different schemas can be used for reading and writing data (White, 2015).

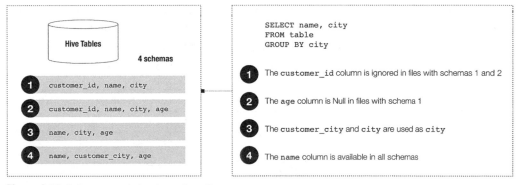

Figure 4-12. Schemas evolution in an Avro file.

4.1.1.6. Parquet

The Parquet format (Parquet, 2018) is an open source columnar file format for Hadoop that uses complex nested data structures to support efficient data compression. Developed initially by Google (Melnik et al., 2010), this approach combines multi-level execution trees with a columnar data layout for efficient data processing (Figure 4-13).

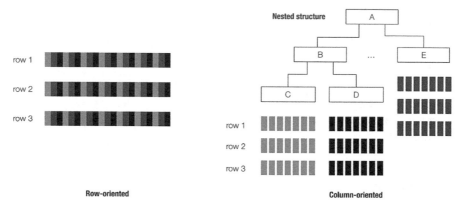

Figure 4-13. The Parquet data format. Adapted from Melnik et al. (2010).

The data structure implemented by the Parquet format allows nested fields to be read independently from other fields, improving performance in query processing and supporting a very efficient compression mechanism. The file format includes a header that is followed by one or more blocks, ending with a footer. The header only includes a 4-byte value, PAR1, identifying the beginning of a Parquet format. All the metadata is stored in the footer, including the format version, the schema, any extra key-value pairs, and the metadata for each specific block. The footer ends with two 4-byte values, including the length of the footer metadata and the PAR1. These files can be split for parallel processing, as blocks can be located right after the footer has been read. Regarding blocks, each block stores a row group containing column chunks with the column data of those rows. Column chunks are written in pages (Figure 4-14) (White, 2015).

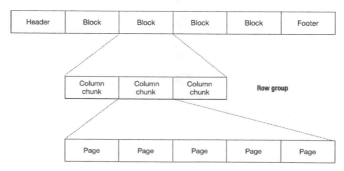

Figure 4-14. Data organization in a Parquet file. Adapted from White (2015).

4.1.2. Partitions and Buckets

For the different file formats, Hive tables can use partitions (and subpartitions) as well as buckets to organize the data, allowing better query performance if the partitions and buckets are defined to accommodate the most common queries. Using once again the customer table as an example, the DDL instruction to create a table using a partition based on the `region` column and a subpartition based on the `nation` column is as follows:

```
CREATE TABLE `customerp`(
  `custkey` string,
  `name` string,
  `address` string,
  `city` string,
  `phone` string,
  `mktsegment` string)
PARTITIONED BY (
  `region` string,
  `nation` string)
ROW FORMAT DELIMITED
  FIELDS TERMINATED BY '\;'
STORED AS INPUTFORMAT
  'org.apache.hadoop.mapred.TextInputFormat'
OUTPUTFORMAT
  'org.apache.hadoop.hive.ql.io.HiveIgnoreKeyTextOutputFormat'
LOCATION
  'hdfs://yourservername:8020/user/lid4/book/hivessb.db/customerp'
```

In this case, the data is organized across various files, distributed along different folders derived from the possible region values and the nations within each region. Figure 4-15 shows an overview of the folders structure using the Ambari interface for browsing the data. The image shows the various regions alongside the nations available in the dataset for a given region (`Europe`).

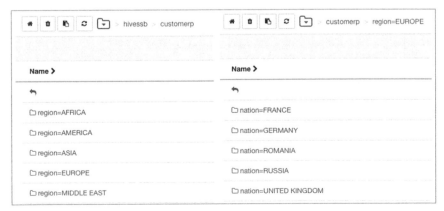

Figure 4-15. Partitions and subpartitions in a Hive table.

The attributes used to define the partitions and subpartitions are not formal columns within the table, as can be seen in the set of instructions for creating the partitioned table, which was previously shown. Nevertheless, those attributes are part of the table and, for this reason, DML instructions can use them to query the data. As shown in Figure 4-16, these attributes are also listed in the results as long as the query does not explicitly exclude them.

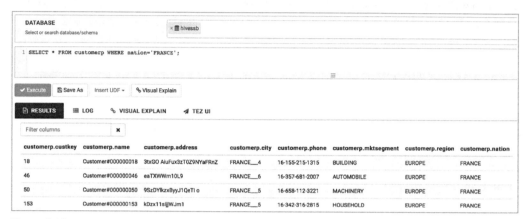

Figure 4-16. Partitions and subpartitions when querying the data.

The use of the command shell to inspect the folders, subfolders, and corresponding files in the partitions is illustrated in the following examples. First, listing the table folder (`customerp`) along its several subfolders; then, the subfolder EUROPE with its various nations as subfolders; followed by the subfolder FRANCE with its file (there are no further data divisions); and lastly, the inspection of a specific file (`000000_0`).

```
$ hdfs dfs -ls /user/lid4/book/hivessb.db/customerp
Found 5 items
drwxr-xr-x   - lid4 0 2018-03-12 20:15 /user/lid4/book/hivessb.db/customerp/region=AFRICA
drwxr-xr-x   - lid4 0 2018-03-12 20:15 /user/lid4/book/hivessb.db/customerp/region=AMERICA
drwxr-xr-x   - lid4 0 2018-03-12 20:15 /user/lid4/book/hivessb.db/customerp/region=ASIA
drwxr-xr-x   - lid4 0 2018-03-12 20:15 /user/lid4/book/hivessb.db/customerp/region=EUROPE
drwxr-xr-x   - lid4 0 2018-03-12 20:15 /user/lid4/book/hivessb.db/customerp/region=MIDDLE EAST
$ hdfs dfs -ls /user/lid4/book/hivessb.db/customerp/region=EUROPE
Found 5 items
drwxr-xr-x   - lid4 0 2018-03-12 20:15 /user/lid4/book/hivessb.db/customerp/region=EUROPE/nation=FRANCE
drwxr-xr-x   - lid4 0 2018-03-12 20:15 /user/lid4/book/hivessb.db/customerp/region=EUROPE/nation=GERMANY
drwxr-xr-x   - lid4 0 2018-03-12 20:15 /user/lid4/book/hivessb.db/customerp/region=EUROPE/nation=ROMANIA
drwxr-xr-x   - lid4 0 2018-03-12 20:15 /user/lid4/book/hivessb.db/customerp/region=EUROPE/nation=RUSSIA
drwxr-xr-x   - lid4 0 2018-03-12 20:15 /user/lid4/book/hivessb.db/customerp/region=EUROPE/nation=UNITED KINGDOM
$ hdfs dfs -ls /user/lid4/book/hivessb.db/customerp/region=EUROPE/nation=FRANCE

-rwxr-xr-x   2 lid4 5151399 2018-03-12 20:15 /user/lid4/book/hivessb.db/customerp/region=EUROPE/nation=FRANCE/000000_0

$ hdfs dfs -text /user/lid4/book/hivessb.db/customerp/region=EUROPE/nation=FRANCE/000000_0
```

```
18;Customer#000000018;3txGO AiuFux3zT0Z9NYaFRnZ;FRANCE___4;16-155-215-1315;BUILDING
46;Customer#000000046;eaTXWWm10L9;FRANCE___6;16-357-681-2007;AUTOMOBILE
50;Customer#000000050;9SzDYlkzxByyJ1QeTI o;FRANCE___5;16-658-112-3221;MACHINERY
153;Customer#000000153;kDzx11sIjjWJm1;FRANCE___5;16-342-316-2815;HOUSEHOLD
204;Customer#000000204;7U7u2KryFP;FRANCE___4;16-761-837-4820;BUILDING
271;Customer#000000271;O,HceV3 XE77Yx;FRANCE___2;16-621-282-5689;MACHINERY
281;Customer#000000281;x5gJ8IGm2Fo9Jigt;FRANCE___0;16-809-382-6446;BUILDING
284;Customer#000000284;2ZgAkaBgb6aigORfIfUd3kHbP;FRANCE___2;16-161-235-2690;AUTOMOBILE
312;Customer#000000312;cH6XucXV0V;FRANCE___1;16-316-482-2555;AUTOMOBILE
319;Customer#000000319; UQ5mF3sdoZT2;FRANCE___3;16-734-928-1642;FURNITURE
402;Customer#000000402;8Cw4p1m1gKYVUgomkAq,es1Zt;FRANCE___1;16-950-729-1638;AUTOMOBILE
413;Customer#000000413;,4Jm5N0ruhJCB7cBR6Kw;FRANCE___5;16-158-285-7336;FURNITURE
431;Customer#000000431;RNfSXbUJkgUlBBPn;FRANCE___3;16-326-904-6643;HOUSEHOLD
452;Customer#000000452;,TI7FdTc gCXUMi09qD;FRANCE___3;16-335-974-9174;BUILDING
455;Customer#000000455;sssuscPJ,ZYQ8viO;FRANCE___5;16-863-225-9454;BUILDING
459;Customer#000000459;CkGH34iK 9vAHXeY7 wAQIzJa;FRANCE___0;16-927-662-8584;MACHINERY
516;Customer#000000516;EJwOQMTQnFwvd8r Y7f9i5POy;FRANCE___1;16-947-309-2690;MACHINERY
539;Customer#000000539;FoGcDu9llpFiB LELF3rdjaiw;FRANCE___6;16-166-785-8571;HOUSEHOLD
    ...
```

Buckets can be defined with or without partitions to reorganize the data in the physical files; they can have a positive or a negative impact on query performance, as they can optimize loading time and execution time when the attribute used for bucketing is also used in the conditions of the queries. Although there is no strict rule to define the optimal number of buckets, it is recommended that a bucket has at least the size of a HDFS block or some multiple of that size (E. Costa, Costa, & Santos, 2018).

To illustrate how buckets are defined, the customer table is once again used. Bucketing clusters the data according to a predefined attribute. This task distributes the data homogeneously throughout the existing buckets. This is the reason why attributes with high cardinality should be considered for bucketing, as this method will concentrate data records instead of spreading them across several smaller files.

Considering a scenario in which four buckets are created for the customer table, based on the `nation` column, the customers are distributed in the four buckets, maintaining a reasonable distribution of the data.

```
CREATE TABLE `customerbn`(
    `custkey` int,
    `name` string,
    `address` string,
    `city` string,
    `nation` string,
    `region` string,
    `phone` string,
    `mktsegment` string)
CLUSTERED BY (
    nation)
INTO 4 BUCKETS
```

In this case, and to understand how bucketing works, the result of the bucketing process is now analyzed. The following SQL statements allow the selection of data from a specific bucket, in order to verify the distribution of the customers according to the nations, by computing the number of rows per nation.

```
SELECT nation, count(*) AS totalb1 FROM hivessb.customerbn
 TABLESAMPLE(BUCKET 1 OUT OF 4 ON nation)
 GROUP BY nation;
```

```
SELECT nation, count(*) AS totalb2 FROM hivessb.customerbn
 TABLESAMPLE(BUCKET 2 OUT OF 4 ON nation)
 GROUP BY nation;
```

```
SELECT nation, count(*) AS totalb3 FROM hivessb.customerbn
 TABLESAMPLE(BUCKET 3 OUT OF 4 ON nation)
 GROUP BY nation;
```

```
SELECT nation, count(*) AS totalb4 FROM hivessb.customerbn
 TABLESAMPLE(BUCKET 4 OUT OF 4 ON nation)
 GROUP BY nation;
```

After running the statements for the four buckets, Figure 4-17 shows the result indicating the various nations and the number of records in each bucket. The number of records is 419,192 (bucket 1), 240,491 (bucket 2), 359,687 (bucket 3) and 480,630 (bucket 4). This type of distribution is useful for performing joins between tables because the attributes used for joining can be grouped in a similar way by the different tables.

nation	totalb1	nation	totalb2	nation	totalb3	nation	totalb4
BRAZIL	59952	ALGERIA	59916	CANADA	59849	ARGENTINA	59841
INDONESIA	60236	ETHIOPIA	60471	IRAN	60101	CHINA	60065
JORDAN	59909	IRAQ	60056	JAPAN	59757	EGYPT	59969
KENYA	59476	ROMANIA	60048	MOZAMBIQUE	59796	FRANCE	60316
MOROCCO	59834			SAUDI ARABIA	59803	GERMANY	60153
PERU	59788			UNITED KINGDOM	60381	INDIA	60215
VIETNAM	59997					RUSSIA	60065
						UNITED STATES	60006

Figure 4-17. Analyzing the distribution of the data with the `nation` bucketing.

The bucket number to which each row should be assigned is identified by the bucketing column using a hash function, so it is influenced by the selected column and its corresponding data type. In the previous example, a `string` column was used, so the bucket number is derived from the value obtained through some computation on the characters within the string. In the case of an `integer` column, the identification of the bucket number is a straightforward process, since it is the result of the function `hash(column) mod [number of buckets]`.

Using the `custkey` column for bucketing, converting the key values stored as `string` values into `integer` values, and using again 4 buckets to segment the data, the distribution of the data is very different from the one previously obtained (using `nation` as the bucketing attribute). When the process is completed, the number of records is 375,000 per bucket. This even distribution of the data is possible thanks to the characteristics of the dataset, which is synthetic and whose records are generated following a specific pattern of replication from the base data.

```
CREATE TABLE `customerb`(
  `custkey` int,
  `name` string,
  `address` string,
  `city` string,
  `nation` string,
  `region` string,
  `phone` string,
  `mktsegment` string)
CLUSTERED BY (`custkey`)
  INTO 4 buckets
INSERT OVERWRITE TABLE hivessb.customerb
  SELECT CAST(custkey AS INT), name, address, city, nation, region, phone,
      mktsegment FROM hivessb.customer;
```

In this case, although the buckets have exactly the same number of rows and all of them include all the nations available in the dataset, the distribution along the different nations and regions is not exactly the same, as shown in Figure 4-18.

nation	totalb1	nation	totalb2	nation	totalb3	nation	totalb4
ALGERIA	14901	ALGERIA	14946	ALGERIA	15063	ALGERIA	15006
ARGENTINA	14917	ARGENTINA	14965	ARGENTINA	15127	ARGENTINA	14832
BRAZIL	14913	BRAZIL	15058	BRAZIL	15014	BRAZIL	14967
CANADA	14953	CANADA	15218	CANADA	15056	CANADA	14622
CHINA	14973	CHINA	14945	CHINA	15064	CHINA	15083
EGYPT	14975	EGYPT	14969	EGYPT	14936	EGYPT	15089
ETHIOPIA	15037	ETHIOPIA	15089	ETHIOPIA	15050	ETHIOPIA	15295
FRANCE	15136	FRANCE	14976	FRANCE	15221	FRANCE	14983
GERMANY	15109	GERMANY	15271	GERMANY	14827	GERMANY	14946
INDIA	14977	INDIA	15033	INDIA	15175	INDIA	15030
INDONESIA	14908	INDONESIA	14978	INDONESIA	15260	INDONESIA	15090
IRAN	15101	IRAN	14902	IRAN	14902	IRAN	15196
IRAQ	14887	IRAQ	14959	IRAQ	15109	IRAQ	15101
JAPAN	15002	JAPAN	14949	JAPAN	14943	JAPAN	14863
JORDAN	15232	JORDAN	14869	JORDAN	14930	JORDAN	14878
KENYA	14849	KENYA	14981	KENYA	14867	KENYA	14779
MOROCCO	15027	MOROCCO	14908	MOROCCO	14829	MOROCCO	15070
MOZAMBIQUE	15050	MOZAMBIQUE	14885	MOZAMBIQUE	14875	MOZAMBIQUE	14986
PERU	15054	PERU	14994	PERU	14944	PERU	14796
ROMANIA	15015	ROMANIA	14947	ROMANIA	14901	ROMANIA	15185
RUSSIA	15018	RUSSIA	14924	RUSSIA	15131	RUSSIA	14992
SAUDI ARABIA	14837	SAUDI ARABIA	15053	SAUDI ARABIA	14876	SAUDI ARABIA	15037
UNITED KINGDOM	15018	UNITED KINGDOM	15216	UNITED KINGDOM	14959	UNITED KINGDOM	15188
UNITED STATES	14967	UNITED STATES	15094	UNITED STATES	14935	UNITED STATES	15010
VIETNAM	15144	VIETNAM	14871	VIETNAM	15006	VIETNAM	14976

region	totalb1	region	totalb2	region	totalb3	region	totalb4
AFRICA	74864	AFRICA	74809	AFRICA	74684	AFRICA	75136
AMERICA	74804	AMERICA	75329	AMERICA	75076	AMERICA	74227
ASIA	75004	ASIA	74776	ASIA	75448	ASIA	75042
EUROPE	75296	EUROPE	75334	EUROPE	75039	EUROPE	75294
MIDDLE EAST	75032	MIDDLE EAST	74752	MIDDLE EAST	74753	MIDDLE EAST	75301

Figure 4-18. Analyzing the distribution of the data with the `custkey` bucketing.

The files can also be inspected using the shell, allowing to verify their contents. In the following example, four files (one per bucket) are listed in the table folder (`customerb`), and two buckets are inspected, namely the ones stored in the `000000_0` file and in the `000001_0` file.

```
$ hdfs dfs -ls /user/lid4/book/hivessb.db/customerb
```
```
Found 4 items
-rwxr-xr-x   2 lid4 37975257 2018-03-12 20:03 /user/lid4/book/hivessb.db/customerb/000000_0
-rwxr-xr-x   2 lid4 37981386 2018-03-12 20:03 /user/lid4/book/hivessb.db/customerb/000001_0
-rwxr-xr-x   2 lid4 37967729 2018-03-12 20:03 /user/lid4/book/hivessb.db/customerb/000002_0
-rwxr-xr-x   2 lid4 37973177 2018-03-12 20:03 /user/lid4/book/hivessb.db/customerb/000003_0
```
```
$ hdfs dfs -text /user/lid4/book/hivessb.db/customerb/000000_0
```
```
938296Customer#000938296kVF6B4ARJxm,YZmhm49NDFLkGINDIA___1INDIAASIA18-431-610-8456BUILDING
844232Customer#000844232dy1T1u4WqvtvhenJxnjAZXk71INDONESIA3INDONESIAASIA19-438-399-6517FURNITURE
844184Customer#000844184oC31zbkosFcAmQ2yMINDONESIA5INDONESIAASIA19-591-932-4673FURNITURE
241704Customer#0002417040kDzyjcpKVv3VIETNAM__4VIETNAMASIA31-652-416-7050FURNITURE
844624Customer#000844624ZmTmS4ynFpvRJ95ae JZvx1FRINDIA___7INDIAASIA18-152-759-8002BUILDING
844168Customer#0008441688jnyZAjtkNcVYrqYkTuINDONESIA4INDONESIAASIA19-696-630-6239MACHINERY
843960Customer#000843960qrNJRQQVXTRydINDONESIA7INDONESIAASIA19-357-221-8775AUTOMOBILE
337884Customer#0003378840Qz,55I2tRVIETNAM__0VIETNAMASIA31-283-606-8508AUTOMOBILE
337772Customer#000337772YdjeohUMojR3PmgSPp1nt37AMVIETNAM__2VIETNAMASIA31-330-219-2336HOUSEHOLD
844716Customer#000844716cV40afO89qbn3BlINDIA___6INDIAASIA18-384-726-9402BUILDING
337708Customer#0003377088TfPUUHpQ7xUVIETNAM__6VIETNAMASIA31-172-553-5905AUTOMOBILE
337700Customer#00033770015McFMtGIZfHlMdOHfSnsKUWHVIETNAM__4VIETNAMASIA31-408-487-7177MACHINERY
843916Customer#000843916kNtAo2e4MmlnYaKINDONESIA0INDONESIAASIA19-192-860-8416FURNITURE
844772Customer#0008447720oxdzEEySOOMyHZ721xzyZdmOINDIA___8INDIAASIA18-310-804-3145FURNITURE
843876Customer#0008438761k6aR8WK,lINDONESIA5INDONESIAASIA19-521-896-1267HOUSEHOLD
585760Customer#000585760ymqcZ1SisIz1IsxmcJOTFmGCmVIETNAM__7VIETNAMASIA31-826-291-2379AUTOMOBILE
…
```
```
$ hdfs dfs -text /user/lid4/book/hivessb.db/customerb/000001_0
```
```
99697Customer#000099697gznfJlqEM536YMLkrXyerT EGYPT____4EGYPTMIDDLE EAST14-279-267-6722FURNITURE
638237Customer#00063823701iQWar2eAS Bu9CANADA___4CANADAAMERICA13-182-296-4100HOUSEHOLD
638405Customer#000638405hyu3lh6S74esamHVPCANADA___5CANADAAMERICA13-293-435-5992BUILDING
638437Customer#000638437,W6oSIbN5sLcuYNwRA5ICANADA___3CANADAAMERICA13-617-152-1396FURNITURE
638445Customer#000638445FgGEoTPxvDBF1HKI,nEi5ymsKCANADA___2CANADAAMERICA13-934-500-9274FURNITURE
638485Customer#000638485CrD jk0CQTwQ2m6eLI27HSfYpCANADA___0CANADAAMERICA13-948-906-1051HOUSEHOLD
728237Customer#0007282371c7LeAQyKX YxAEYaIUZWCANADA___3CANADAAMERICA13-359-484-5191MACHINERY
728433Customer#000728433pTXe5oMTlp13GK2XYyCANADA___5CANADAAMERICA13-291-993-9522AUTOMOBILE
728465Customer#000728465LRKq0PPgdL,vqYFfJ4VAkH6JQCANADA___2CANADAAMERICA13-180-7311824AUTOMOBILE
728621Customer#000728621TR91,MFEfcoXCANADA___4CANADAAMERICA13-879-335-7798BUILDING
728749Customer#000728749djqfa19A31IhOfvWUFyJ9hgMiCANADA___1CANADAAMERICA13-408-552-3228FURNITURE
728909Customer#000728909NpUGPh2xKOWxnc3Vc9S oophdCANADA___7CANADAAMERICA13-829-655-4740MACHINERY
728945Customer#000728945ix1wtUO7K4a4aHNv,76F,T8YLCANADA___3CANADAAMERICA13-828-461-6602HOUSEHOLD
728981Customer#000728981m32u6ogVHTF,xRplwip, ECaaCANADA___1CANADAAMERICA13-946-7541989AUTOMOBILE
546645Customer#000546645W62WN81CBH60syg,2pC3xNsytCANADA___2CANADAAMERICA13-908-326-3171BUILDING
546673Customer#000546673U4UOtjbLhBxNz20BMy4wnIa89CANADA___7CANADAAMERICA13-501-100-9099MACHINERY
…
```

As will be shown in Chapter 8, partitioning can play an important role in the organization of a Hive table when the different design patterns are benchmarked. Regarding the processing time needed to run a query, the attributes normally used in the "`where`" clause of a query are the most appropriate ones for partitioning and/or subpartitioning.

Bucketing is advantageous for sampling when the volume of data imposes processing limitations. In these cases, data analyzes may occur on a sample of the data. This sample can be obtained through the bucket table sampling strategy, which is optimized for this type of tables. Another advantage of bucketing is the use of bucket map joins in Hive. This is an appropriate strategy for joining large tables bucketed by the join attribute, as long as the number of buckets in one table is a multiple of the number of buckets in the other table (Du, 2015).

4.2. FROM DIMENSIONAL MODELS TO TABULAR MODELS

Business Intelligence and Analytics (BI&A) in Big Data environments can be enhanced with appropriate data systems. In those contexts, the volume, variety, or velocity of data can justify the adoption of different databases, such as NoSQL, as storage repositories supporting operational or analytical applications (H. Chen et al., 2012).

Regarding analytical applications, the Dimensional Data Model (DDM) is used widely to model data requirements that guide the implementation of a DW using traditional environments such as relational or multidimensional databases. However, when the data volume justifies the implementation of a DW in a Big Data context, a data model also needs to be identified to define the structure of the tables in Hive. Due to Hive's characteristics, the data schemas must be defined taking into account the queries that need to be answered, specifying the analytical requirements.

Dimensional modeling is widely used in traditional BI environments, and many DWs are already available to support organizational decision-making processes. As previously stated, this book addresses the definition of an automatic process for transforming a DDM into tables that can be implemented in Hive. Following this, the definition of the analytical requirements for a BDW can be achieved by adopting common and well-known practices of the BI community, such as the DDM. An advantage of this approach is that the elicitation process for identifying the BDW data requirements can follow a combination of a goal-driven, a user-driven, or a data-driven approach (Vieira, Pedro, Santos, Fernandes, & Dias, 2019). This combination ensures that no important data requirements are left behind; and, furthermore, an integrated representation of the data model, the DDM, warrants a common understanding of the organizational business processes, business indicators, and dimensions of analysis.

In case a DDM exists, the guidelines proposed in this section may be followed. If that is not the case, or if organizations and practitioners want to look into the BDW in a global perspective, Chapter 5 addresses the design and implementation of BDWs with an approach where logical components and data flows, technological infrastructures, and data modeling are seen as being part of an integrated process.

In the transformation process described below, a DDM is regarded as an Analytical Data Model (ADM), which can be implemented in a traditional BI environment (ADM$_{BI}$), or in a Big Data environment (ADM$_{BD}$). In a traditional BI environment, relational databases are usually used as storage technologies, whereas in a Big Data environment, Hive is the most common DWing storage technology.

For identifying a possible data model for a BDW, the definition of ADM$_{BI}$ is presented and the definition of ADM$_{BD}$ is proposed based on tabular data models (Santos & Costa, 2016b).

Definition 1 – An ADM in a BI context, $ADM_{BI} = (D, F, R, M)$, includes a set of dimension tables, $D = \{D_1, D_2, ..., D_n\}$, a set of fact tables, $F = \{F_1, F_2, ..., F_n\}$, a set of relationships, $R = \{R_{F_1}^{D_1}(c_f : c_d), R_{F_1}^{D_2}(c_f : c_d), ..., R_{F_m}^{D_i}(c_f : c_d)\}$, associating fact tables to dimension tables. The cardinality of the relationship $(c_f : c_d)$ of n:1 stands between a fact table and the dimension tables. An ADM$_{BI}$ includes a set of measures for each fact table, $M_j = \{M_j^1, ..., M_j^k\}$, used for analyzing the business processes. An abstract example of an ADM$_{BI}$ is provided in Figure 4-19.

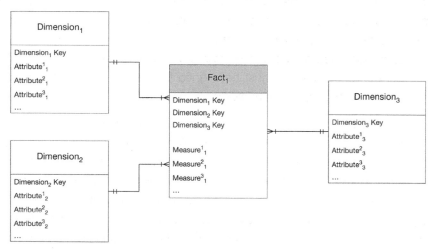

Figure 4-19. Example of an ADM$_{BI}$.

Definition 2 – An ADM in a Big Data context for Hive, $ADM_{BD} = (T, C, A)$, includes a set of tables, $T = \{T_1, T_2, ..., T_n\}$, where each table integrates one or more components, as $T_i = \{C_1, C_2, ..., C_n\}$, which can be descriptive or analytical. Each component integrates different columns representing the attributes with the atomic values that will be stored in the Hive tables, $C_j = \{A_j^1,...,A_j^k\}$, as shown in Figure 4-20.

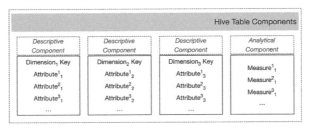

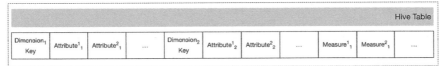

Figure 4-20. Components of a Hive table.

The following subsections describe a set of guidelines defined as rules that can be used to transform a DDM into an ADM_{BD} using an ADM_{BI} as input. We begin by proposing the identification of primary data tables based on raw data. These tables are organized in such a way that a few highly denormalized tables are obtained. After that, derived data tables with aggregates from the raw data can be identified, reducing the number of data records to be analyzed. These derived data tables can then be implemented as materialized views that provide *ad hoc* querying capabilities.

In this process, the identification of tables (either primary or derived) is done after identifying the descriptive components, including the descriptive attributes that will add meaning to the business measures under analysis (analytical components with analytical attributes).

4.2.1. Primary Data Tables

The identification of the primary data tables and the associated components is achieved through the following set of rules.

Rule ADM_{BD}.1 – Identification of primary data tables. The identification of primary data tables begins with the fact tables present in the ADM_{BI}. Each fact table originates a primary data table, integrating the raw data that will allow the computation of queries in an ADM_{BD}.

Figure 4-21 shows a constellation schema with two fact tables; two primary data tables are identified: `Fact`$_1$ and `Fact`$_2$.

Rule ADM_{BD}.2 – Identification of components. The identification of components of an ADM_{BD} follows a two-step approach; one for the descriptive components and one for the analytical components.

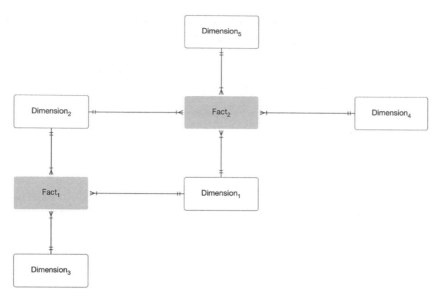

Figure 4-21. Example of a data model with a constellation schema.

Rule ADM$_{BD}$.2.1 – Identification of descriptive components. All the dimensions present in the ADM$_{BI}$ will lead to descriptive components, one per dimension, providing in turn the descriptive attributes that add semantics to the analytical components, representing the different analytical perspectives in a DDM. The attributes of a descriptive component include all the attributes present in the corresponding dimension of the ADM$_{BI}$. Following the best practices for designing a DDM, the dimensions' key must be a durable surrogate key (SK), which belongs to the DW and not to the original data sources, thus ensuring the consistency of the DW. In the case of a BDW, these keys may or may not exist depending on how the BDW updates are foreseen, in other words, depending on the types of Slowly Changing Dimensions (SCDs) to be implemented.

Showing an abstract example and also a specific example associated with the `Calendar` dimension, Figure 4-22 shows the identification of the corresponding descriptive components. In this example, the `Calendar` dimension includes a SK, which is the PK of the dimension and a natural key (NK) that helps to implement SCD Type 2 within dimensions that require this type of SCD.

Rule ADM$_{BD}$.2.2 – Identification of analytical components. All the fact tables present in the ADM$_{BI}$ lead to analytical components, one per fact table, providing the analytical attributes that will be analyzed according to the related descriptive components. The attributes of an analytical component are composed of the set of non-key attributes (excluding all PKs and/or FKs) present in the corresponding fact table within the ADM$_{BI}$. Those attributes are the measures or business indicators considered in the DDM.

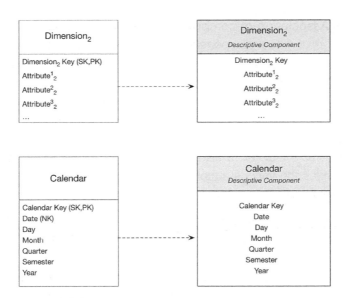

Figure 4-22. Example of a descriptive component.

Figure 4-23 shows two examples where the analytical components are identified for both an abstract example and for the `Sales` fact table. All key attributes are ignored, as the descriptive components will include the necessary information regarding the dimensions' SKs.

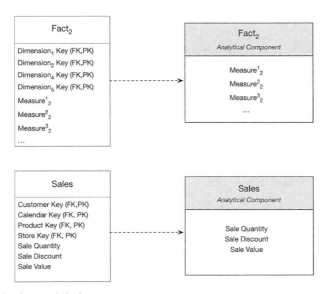

Figure 4-23. Example of an analytical component.

Rule ADM$_{BD}$.3 – Association of components to primary tables. A primary data table includes a set of descriptive components and one analytical component, which vary depending on the table. For identifying the descriptive components and the analytical component, the DDM must be split into star schemas, when dealing with constellation schemas, in order to denormalize each star schema. After each star schema is identified, the primary data table will treat all the components derived from the association between dimensions and the fact table as descriptive. The analytical component is the one derived from the star schema's fact table. Each one of these components, either descriptive or analytical, provides the various attributes (columns) that must be considered for the physical implementation of the Hive tables.

Returning to the constellation schema previously shown in Figure 4-21, two star schemas integrate this constellation and, for each one, a primary data table is derived. Figure 4-24 shows the denormalization process that must be followed. Although this is a straightforward denormalization, these steps are afterwards extended for identifying the derived data tables (subsection 4.2.2), showing how they aggregate data, and how they impact data volume and query performance (section 4.3).

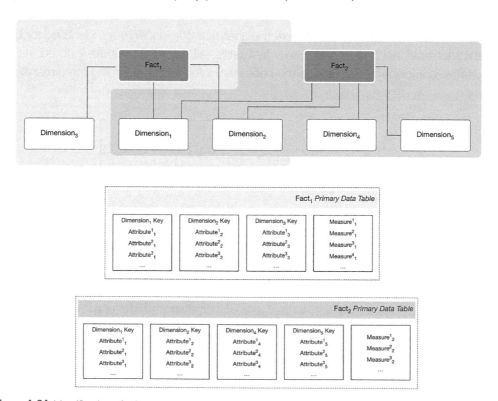

Figure 4-24. Identification of primary data tables.

4.2.2. Derived Data Tables

After identifying the primary data tables, derived data tables can be generated to group descriptive and analytical components. It is important to bear in mind that when they are grouped according to a descriptive component, analytical components are aggregated/summarized using aggregation functions. This means that derived tables are optimized to answer specific queries, as they pre-compute aggregates on data, hence decreasing the number of available records. Although primary data tables can answer any query, at any level of detail, as long as the granularity of the model complies with them, derived data tables do not, as pre-aggregations of data are computed for optimizing performance.

The following rules can be observed for identifying the derived data tables and their components.

Rule ADM$_{BD}$.4 – Identification of derived data tables. The identification of derived data tables considers the dimension tables present in the ADM$_{BI}$, and computes the dimensions lattice to verify the different combinations of dimensions. This lattice will identify analytical contexts where different fact tables can be integrated in the same derived data table, as those fact tables share a subset of dimensions defining an analytical context. Moreover, for a specific fact table, different levels of detail determine different analytical contexts.

Considering an abstract example in which a star schema integrates a fact table (Fact$_1$) and three dimensions (Dimension$_1$, Dimension$_2$, and Dimension$_3$) as shown in Figure 4-25 (top), the dimensions lattice, shown in Figure 4-25 (bottom), integrates the seven possible tables with all the combinations between dimensions. This lattice includes the primary data table (gray rectangle) and the six derived data tables with aggregated data, at different levels of detail, providing different analytical perspectives on the data. This lattice excludes the level of detail where a single row summarizes all the data, as can be seen later in section 4.3.

When considering a constellation schema, that is when more than one fact table is available in the data model, the number of tables must take into account the combination of dimensions as well as the fact tables to which they are related, since fact tables can share different dimensions. This is an important point to bear in mind, because this will allow to merge different fact tables with complementary analytical measures into the same derived table.

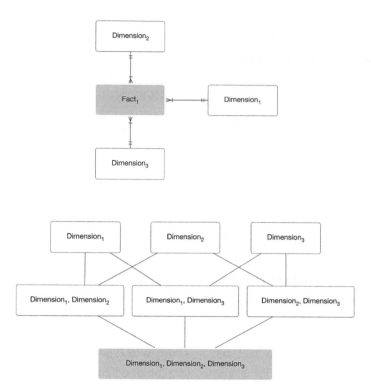

Figure 4-25. Example of a dimensions lattice of a star schema: ADM_{BI} example data model (top) and dimensions lattice (bottom).

Considering a data model with a constellation schema, as shown in Figure 4-26, where two fact tables (`Fact₁` and `Fact₂`) share two dimension tables (`Dimension₁` and `Dimension₂`), the number of possible combinations that can be obtained is 15 (as shown in Figure 4-27), however, the total number of derived tables is lower. Since fact tables usually do not share all the available dimension tables in a data model, combinations that propose integrating dimensions that are not shared by the fact tables must be removed.

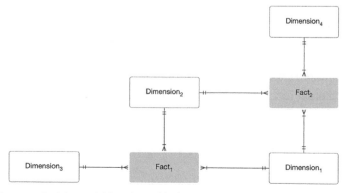

Figure 4-26. ADM_{BI} example data model (two fact tables).

Figure 4-27 shows the combinations that must be excluded as dashed rectangles, whereas the two primary data tables are represented by gray rectangles. In this example, all combinations including `Dimension₃` and `Dimension₄` must be excluded from the final set of derived tables, since they are not shared by the two fact tables.

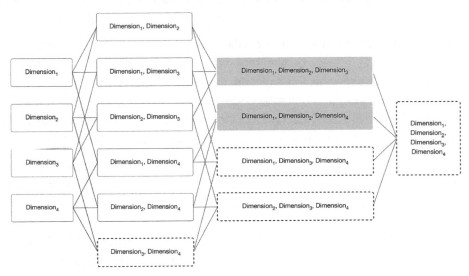

Figure 4-27. Dimensions lattice for a constellation schema.

After removing these impossible combinations, 11 different tables are effectively identified, as shown in Figure 4-28; 2 are associated with the primary data tables and the remaining nine with the derived data tables at different levels of detail.

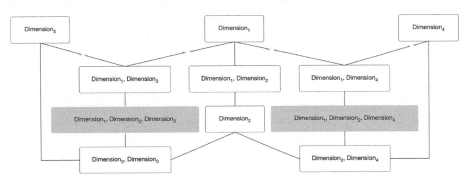

Figure 4-28. Valid combinations of a dimensions lattice for a constellation schema.

After a set of valid derived tables is identified, it is possible to continue assigning the descriptive and analytical components to each table. To do so, the following rules must be followed.

Rule ADM$_{BD}$.5 – Association of components to derived tables. A derived data table integrates a set of descriptive components and one or more analytical components.

Rule ADM$_{BD}$.5.1 – Descriptive components. Taking into account the dimensions lattice identified in rule ADM$_{BD}$.4, each derived table will integrate the descriptive components of the corresponding dimensions. For example, the table integrating data from Dimension$_1$ and Dimension$_3$ will include the corresponding descriptive components from these two dimensions, whereas the table that only includes data from Dimension$_3$ will integrate only this descriptive component. This process must be followed for all derived tables.

Rule ADM$_{BD}$.5.2 – Analytical components. To identify the analytical components, it is useful to analyze the corresponding ADM$_{BI}$ to check the fact tables associated to each derived table. Considering the existing relationships between the fact tables and the dimension tables of the ADM$_{BI}$, each table in the ADM$_{BD}$ will include as analytical components those inherited from the relationships expressed in the ADM$_{BI}$. For example, in Figure 4-26, the table derived from Dimension$_1$ will include the analytical components derived from Fact$_1$ and from Fact$_2$, whereas the table derived from Dimension$_2$ and Dimension$_4$ will only include the analytical component derived from Fact$_2$.

As was previously mentioned, all these derived tables include data at a particular level of detail, hence the reason why they need aggregation functions for the attributes present in the analytical components, as different measures summaries need to be calculated, according to the descriptive components present in a specific derived table. By default, it is assumed that the aggregation function is SUM, although other functions can be used, for instance, AVG, COUNT, MIN, or MAX.

In order to guide the assignment of the descriptive and analytical components to the primary and derived data tables identified by the rules already described in this section, it is advisable to identify the data matrix that summarizes the association between them. This prevents potential mistakes from happening when the number of tables increases the complexity of this task. Considering the example previously shown in Figure 4-26, Table 4.1 lists the resulting data matrix, which assigns to each derived table its corresponding descriptive and analytical components. The matrix also includes the primary data tables, providing an overall overview of all tables.

After identifying all primary or derived tables, as well as their corresponding descriptive or analytical components, the modeling of the DW can proceed with the identification of partitions and buckets.

Table 4.1. Data matrix with the identification of the descriptive (O) and analytical (⊙) components.

	Dimension$_1$	Dimension$_2$	Dimension$_3$	Dimension$_4$	Fact$_1$	Fact$_2$
Derived$_1$	O				⊙	⊙
Derived$_2$		O			⊙	⊙
Derived$_3$			O		⊙	
Derived$_4$				O		⊙
Derived$_5$	O	O			⊙	⊙
Derived$_6$	O		O		⊙	
Derived$_7$	O			O		⊙
Derived$_8$		O	O		⊙	
Derived$_9$		O		O		⊙
Primary$_1$	O	O	O		⊙	
Primary$_2$	O	O		O		⊙

Rule ADM$_{BD}$.6 – Identification of partitions and buckets. The identification of partitions and buckets will consider the necessary balance between the cardinality of the attributes and their distribution. For each descriptive attribute, its cardinality and distribution must be analyzed. Since a table in an ADM$_{BD}$ can integrate one or more descriptive components, it is necessary to analyze the cardinality and distribution of all the columns, to identify which ones should be partitioned or bucketed.

The attribute that defines the granularity of the descriptive component is the one that provides the highest level of detail to the table, so the cardinality and distribution of those attributes must be analyzed. Considering different descriptive columns and their characteristics in terms of cardinality and distribution, a column with lower cardinality and uniform distribution should be used as the partitioning attribute, whereas a column with higher cardinality should be used as the bucketing attribute. These suggestions aim to balance the data and enhance query performance. Nevertheless, the suggested model can be fine-tuned by the user or database modeler to add other attributes to the partitions or to adapt these suggestions to the required queries.

This rule aims to be as general as possible, although it is suggested that the Calendar dimension (or Time or a combination of both Time/Calendar) be used as the partition component taking into consideration the hierarchy of its attributes, following a partition path from the most aggregated level to the most detailed one. Afterwards, the partitions can be complemented with a bucketing attribute.

Considering the Calendar dimension or a descriptive component with the attributes shown in Figure 4-29, several hierarchies including these attributes can be defined. Using the one that includes all the available attributes, from the most aggregated one to the most detailed one, the partitions can be defined using Year as the first division, followed by `Semester`, `Quarter`, `Month`, and, finally, `Day`. This is usually a useful approach since many analytical queries include calendar or time restrictions, and thus benefit from this kind of data organization strategy.

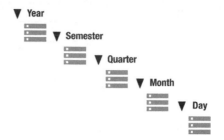

Figure 4-29. Example of partitions for the `Calendar` dimension.

Figure 4-30 presents a star schema with the `Sales` fact table linked to four dimension tables, namely `Calendar`, `Customer`, `Product`, and `Store`. Assuming a scenario where data from 30 years are stored, and considering that the organization has 50,000 products, 10,000 customers, and 500 stores, if 10% of the customers buy per day 0.1% of the products in each store, the fact table will have more than 273 billion records for these years. Instead of having all the data in the same Hive partition, it can be split into approximately 10,950 partitions (as shown in Figure 4-29) for 30 years, meaning that each partition would have an average value of aproximately 25 million records.

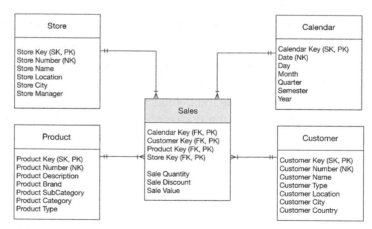

Figure 4-30. Example of the `Sales` star schema.

To clarify how the number of records was estimated for the proposed scenario, the following equation considers all the dimensions and provides the number of facts (rows) for the considered years:

(30 x 365)`Calendar` x 1000`Customers` x 50`Products` x 500`Stores` = 273,750,000,000 records

Following the same line of reasoning, and considering the `Calendar` attributes, the following equation shows how to estimate the number of partitions for splitting the data:

30`Years` x 365`Days` = 10,950 partitions

In a context where such a level of detail is not required, a different distribution of the files and rows can be achieved considering, for instance, a monthly distribution:

30`Years` x 12`Months` = 360 partitions

With 10,950 partitions, each partition will include an average value of 25 million records, whereas for 360 partitions, each one is expected to have an average value of 760 million records.

The organization of the Hive files, in the defined partitions, has an important impact in the time needed for processing queries; it is thus possible to significantly improve performance when partitions are properly defined. To better explain this impact, Chapter 8 offers a benchmark comparing the performance of several queries with and without partitions.

4.3. OPTIMIZING OLAP WORKLOADS WITH DRUID

One of the main advantages of OLAP workloads is its capability to perform *ad hoc* querying on a dataset, thus providing a wide range of queries for analyzing the data. In many cases, if the data model is properly defined, a user can analyze the data from many different perspectives through a user-friendly interface that supports decision-makers in analytical tasks (Correia, Santos, Costa, & Andrade, 2018).

OLAP tools support the interactive process of creating, managing, analyzing, and reporting data, allowing users to analyze large quantities of data in real-time. In this context, data is perceived and manipulated as if it is stored in a multidimensional array. In OLAP, data cubes allow data to be modeled and viewed in multiple dimensions, enabling the analysis of various measures (present in the fact table) through the considered dimensions (Han et al., 2012).

Traditional BI systems benefit from this interactive analytical environment, using the DDM as input for OLAP servers. This *ad hoc* analytical capability is not common in BDWs, since data models in Hive follow, in most cases, a use case driven approach, where tables are defined according to the queries that need to be answered. The *ad hoc* approach means that new tables can be defined and added to the BDW as new analytical tasks emerge.

© FCA

Given the advantages of dimensional models, which are normally used to implement DWs in traditional BI environments, we can see that extending them to a Big Data context allow us to benefit from OLAP-based queries for BDWs.

The previous section discussed a set of rules for transforming dimensional models into suitable ADMs for Hive (ADM_{BD}). Although those rules help practitioners to define Hive models for BDWs, the number of identified tables is usually too high, particularly those associated with derived data tables, because it depends on the number of dimensions of the model.

Given the potentially high number of tables, the question is to decide which tables need to be implemented or materialized following the concept of OLAP cube and the *ad hoc* querying capabilities it provides. For understanding how much impact the level of detail has when analyzing the data, it is essential to recall the concept of cube and the related cuboids, a concept similar to the dimensions lattice discussed in subsection 4.2.2.

In a lattice of cuboids, the cuboid that has the lowest level of summarization is the **base cuboid**. The 0-D cuboid, which includes the highest level of summarization, is the apex cuboid, usually denoted by all (Han et al., 2012). Between the base (n-D) cuboid, and the apex (0-D) cuboid, the remaining cuboids are identified as members of level n_{-1}-D, n_{-2}-D, and so on, as shown in Figure 4-31, where n is the number of dimensions.

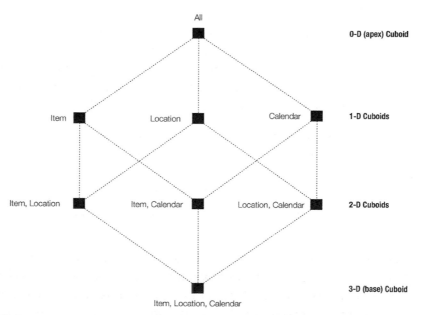

Figure 4-31. Example of a lattice of cuboids for a data cube with three dimensions.

The lattice of cuboids represents all the possible combinations of facts available in a fact table, aggregated by the associated dimensions. In the base cuboid, detailed data provides the highest level of detail, whereas in the apex cuboid, a row aggregates all the data and provides the lowest level of detail. The intermediary cuboids provide different summaries of the data, which are influenced by the granularity of the data and the considered dimensions.

To highlight the tradeoff between the level of detail and the volume of data, which affects the capability to answer different queries, Figure 4-32 shows the effect of data granularity on the storage needs and the querying capability. When data granularity rises, and hence detail is lost, the capacity to answer queries decreases because data is more aggregated. The user normally benefits from lower granularities since they offer the highest levels of detail and ensure more analytical support for decision-making processes.

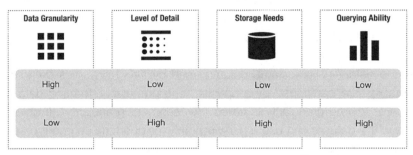

Figure 4-32. Data granularity, level of detail, storage needs, and querying ability. Adapted from Correia et al. (2018).

In a traditional OLAP environment, when a multidimensional database is used, the OLAP server can materialize all cuboids (full materialization), materialize a subset of the possible cuboids (partial materialization) or not pre-compute any non-base cuboid (no materialization) (Han et al., 2012). In the last case, the processing engine can be slow when answering queries because the aggregates over the data are computed as users require them. In the opposite circumstance, all aggregates are pre-computed and thus require huge amounts of storage capacity.

When moving to a Big Data context, the full materialization of a cube is possible because storage capacity and processing capabilities are not usually a problem. Nevertheless, in a modest scenario where a fact table is linked to 10 dimension tables, the number of cuboids can be calculated as 2^{10}, i.e., 1024 different tables. This number excludes the hierarchies normally included in the dimension tables so, in a real scenario, the number of possible cuboids is considerably higher.

Materializing some views plays an important role in increasing the performance of data processing, therefore, it is essential to identify which views should be materialized. An alternative to view materialization is provided by some specific types of databases, such as Druid, which can store and organize data while providing online analytical capabilities over data. This is achieved through the time (and/or calendar) dimension that organizes the data (Yang et al., 2014).

Druid (http://druid.io) is a column-oriented data store intended for sub-second *ad hoc* interactive queries for grouping, filtering, and aggregating data. It is designed for real-time exploratory analytics of large datasets. Its efficiency is enhanced by an advanced indexing structure and a low latency data ingestion. Tables in Druid are called "data sources" and include time-stamped events, usually partitioned into segments. Segments typically include a collection of 5 to 10 million rows spanning over a period of time and are the fundamental organization unit in terms of storage, distributing the data (Yang et al., 2014).

Druid uses a timestamp column to distribute data along the segments, which is also useful for implementing data retention policies and query pruning. The time interval required for partitioning the data depends on the data volume and its time range, since a data source can be partitioned by hour, day, week, month, or another time frame that adequately distributes the data. The timestamp column used by Druid simplifies data distribution, as Druid's tables are partitioned by a specific time granularity, forming data segments or "data shards". Time granularity is called the "segment granularity". If data is spread across a year, the dataset can use a daily, monthly, or quarterly segment granularity, for instance. If timestamps are spread across a day, the dataset is better partitioned by hour, for example (Correia et al., 2018). Segments are immutable blocks of data allowing data consistency and data concurrency, so multiple users can simultaneously use the same data.

An example of the application of a daily segment granularity can be seen in Figure 4-33. In this case, the data is distributed according to different data segments that enhance performance, hence only the required segments are accessed for data processing.

Segments in Druid are identified by a data source identifier, the time interval of the data, and a versioning string (which increases when a new segment is created). This versioning string shows where the new data is, since later versions have the more recent data. This metadata is used by the system for concurrency control, ensuring that read operations always access data of a specific time range from the latest version (Yang et al., 2014). The next listing code shows a JSON object that includes a segment's metadata (for a SSB table, with a SF of 300, detailed later in this section) partitioned in the time frame by Quarter, which includes information about the data source itself (the dimensions and metrics available in the dataset), the time interval of this specific segment, its size, location, and identifier.

Figure 4-33. Example of data segments in Druid.

```
{"dataSource":"analytical_obj300_SQuarter",
 "interval":"1992-01-01T00:00:00.000Z/1992-04-01T00:00:00.000Z",
 "version":"2018-07-03T23:46:51.020Z",
 "loadSpec":{"type":"hdfs",
  "path":"hdfs://host:port/apps/druid/warehouse/analytical_obj300_SQuarter/
  19920101T000000.000Z_19920401T000000.000Z/2018-07-03T23_46_51.020Z/0_index.zip"},
 "dimensions":"c_city, c_nation, c_region, s_city, s_nation, s_region, p_mfgr,
             p_category, p_brand1, od_year, od_yearmonthnum, od_yearmonth",
 "metrics":"quantity, extendedprice, discount, revenue, supplycost",
 "shardSpec":{"type":"none"},
 "binaryVersion":9,
 "size":1337536085,
 "identifier":"analytical_obj300_SQuarter_1992-01-01T00:00:00.000Z_1992-04-
             01T00:00:00.000Z_2018-07-03T23:46:51.020Z"}
```

Besides time, other attributes may be used to further partition the data and increase performance. This may be needed to match the optimal segment's size. Druid's documentation refers the optimal segment's size as including around 5 million records and a storage space between 500MB and 1GB (Yang et al., 2014). After the segments are defined, two types of partitioning can be used: hashed partitions (based on the hash of all the dimensions in each row) or dimension partitions (based on value ranges for a single dimension). Druid's documentation recommends hashed partitions over dimension partitions, since in most cases the former improves indexing performance and creates more homogeneous data segments (Druid, 2018). As can be seen in Figure 4-34, following the previous example for identifying data segments, defining three shards (hashed partitions) will distribute the data of each segment across three files with an approximate data volume.

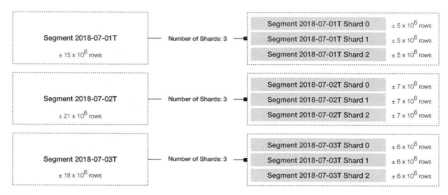

Figure 4-34. Definition of data shards in Druid.

Distributing the data across various segments or shards balances the size of the files, but also provides a first level of data pruning when querying the data, since only the required segments or shards are processed. Druid also allows the use of query granularity during the former process. When using query granularity, Druid applies a data aggregation mechanism (a data roll-up) during data ingestion which compacts the raw data according to a predefined level of detail. Consequently, the volume of data is reduced, since similar records are grouped together. Nonetheless, this method reduces the capacity to carry out more detailed analyzes, because certain details about the aggregated data are lost during grouping. Query granularity has a strong impact on the overall performance, since reducing data volume usually increases query performance, but may compromise some data analysis tasks, due to the reasons previously given. Figure 4-35 shows the impact of applying a daily query granularity during data ingestion. Details about the hours are lost; therefore, in the future, this level of detail cannot be used when querying the data.

Since using query granularity in Druid allows creating materialized views of the data, these are tables that are useful for answering frequently repeated queries. To understand the impact of the different organization strategies for analyzing data, an example is now presented using data segments, data shards (for hashed partitions), and query granularity.

Timestamp	Product Description	Order Value
2018-07-01T10:00:00Z	Product X	€125
2018-07-01T11:00:00Z	Product Y	€110
2018-07-01T12:00:00Z	Product X	€115
2018-07-01T13:00:00Z	Product X	€100
2018-07-02T09:00:00Z	Product Y	€225
2018-07-02T10:00:00Z	Product Y	€125
2018-07-02T11:00:00Z	Product X	€100
2018-07-02T13:00:00Z	Product X	€125
2018-07-03T10:00:00Z	Product X	€155
2018-07-03T11:00:00Z	Product Y	€125
2018-07-03T12:00:00Z	Product X	€125
2018-07-03T13:00:00Z	Product Y	€155

Query Granularity: Day

Timestamp	Product Description	Order Value
2018-07-01T	Product X	€340
2018-07-01T	Product Y	€110
2018-07-02T	Product X	€225
2018-07-02T	Product Y	€350
2018-07-03T	Product X	€280
2018-07-03T	Product Y	€280

Figure 4-35. Definition of query granularity in Druid.

For querying the SSB (a data model introduced in section 4.1), 13 different queries are available (O'Neil et al., 2009). In this subsection, a subset of these queries is highlighted, using a denormalized version of the original SSB queries (because Druid does not support joins). Druid's performance with all of them was already tested and the detailed results are available in Correia et al. (2018). To illustrate Druid's performance with different data organization strategies, three queries from the second query group are used:

```
Q2.1
SELECT SUM(REVENUE), ORDERDATE_YEAR, P_BRAND
FROM LINEORDER_FLAT
WHERE P_CATEGORY = 'MFGR#12' AND S_REGION = 'AMERICA'
GROUP BY ORDERDATE_YEAR, P_BRAND
ORDER BY ORDERDATE_YEAR, P_BRAND


Q2.2
SELECT SUM(REVENUE), ORDERDATE_YEAR, P_BRAND
FROM LINEORDER_FLAT
WHERE P_BRAND BETWEEN 'MFGR#2221' AND 'MFGR#2228 AND S_REGION = 'ASIA'
GROUP BY ORDERDATE_YEAR, P_BRAND
ORDER BY ORDERDATE_YEAR, P_BRAND


Q2.3
SELECT SUM(REVENUE), ORDERDATE_YEAR, P_BRAND
FROM LINEORDER_FLAT
WHERE P_BRAND= 'MFGR#2239' AND S_REGION = 'EUROPE'
GROUP BY ORDERDATE_YEAR, P_BRAND
ORDER BY ORDERDATE_YEAR, P_BRAND
```

© FCA

This example uses a SF of 300, meaning that the "`lineorder`" table has around 18 x 10^6 rows for seven years of data. Table 4.2 shows the number of created segments, the needed storage space, and the average number of rows per segment using two segment granularities, `Month` and `Quarter`. Here, the `Day` segment's granularity is omitted because it would require more than 2400 small-sized segments for the seven years, and such outcome would normally have a negative impact on performance (Correia et al., 2018). `SMonth` (segment by month) or `SQuarter` (segment by quarter) provide less detail, but also a lower number of segments and, therefore, larger file's sizes. When executing queries (`Q2.1`, `Q2.2`, and `Q2.3`), the impact of each granularity can be seen in the time required to process the data. As also shown in Table 4.2, in one of the heaviest groups of queries from the SSB, the segments using the monthly granularity perform better. Readers interested in the performance of Druid using other queries and SFs, where a time reduction of up to 80% can be achieved by selecting a specific segment granularity, should see (Correia et al., 2018).

Table 4.2. Example of different segment's granularities in Druid. Adapted from Correia et al. (2018).

Table	Number of Segments	Average Size per Segment (MB)	Average Rows per Segment (x 10^6)	Q2.1 (sec)	Q2.2 (sec)	Q2.3 (sec)
SMonth	80	1060	22.50	4.63	1.57	0.98
SQuarter	27	3150	66.67	5.21	2.07	1.01

Another important feature of Druid is its ability to cache the results of previous queries, reusing previously computed results and hence substantially decreasing query response time. In Druid, when a query is received it is automatically mapped to a set of segments. If the results already exist in the system cache, no recomputation is done. If the results are not available yet, the query is processed using the corresponding historical or real-time data. In the case of historical data, the results will be cached in the segment history for later use. For real-time data, no caching approach is implemented, because this data is continuously changing. Furthermore, Druid uses a memory-mapped storage engine, which relies on the operating system to page segments in and out of memory. Given that segments can only be scanned if they are loaded into memory, recent segments are retained in memory whereas segments that have never been queried are paged out of memory (Yang et al., 2014). In this way, newer queries can benefit if they need segments used by previous queries. Druid uses HDFS as a deep storage component to store data indexes. Druid's historical nodes download the data segments from HDFS, before querying the data.

The effects of Druid's cache mechanism are visible in Table 4.2 for `Q2.1`, `Q2.2`, and `Q2.3`. As previously shown in the listing code, these three queries are very similar. Nevertheless, `Q2.1` needs more time to be computed, since this is the first query of this group computed by the system. Once its results are computed, `Q2.2` and `Q2.3` benefit from the cache, reusing the results previously computed by previous similar queries. This is one of Druid's most important characteristics because it significantly increases the overall performance of an analytical system.

To better understand this cache mechanism, these results are now analyzed in more detail for SMonth. Listed in Table 4.2 is the average time obtained after four runs of each query. Looking at the individual time it took for each query to run, the results in Table 4.3 show that Run 2 of Q2.1 required 45% less time than Run 1, whereas Run 4 required 81% less time than Run 1. The effect of the cache is also visible in Q2.2 and Q2.3. In these cases, Run 2, Run 3, and Run 4 required less time than Run 1, but also Run 1 from Q2.2 required 76% less time than Run 1 of Q2.1; whereas Run 1 of Q2.3 required 89% less time than Run 1 of Q2.1. The effect of the cache is propagated throughout the various runs, substantially decreasing the time needed to obtain the results.

Table 4.3. Example of the cache mechanism in Druid (time in seconds). Adapted from Correia (2018).

Query	Run 1	Run 2	Run 3	Run 4	Average
Q2.1	8.76	4.83	3.26	1.68	4.63
Q2.2	2.12	1.28	1.28	1.62	1.57
Q2.3	0.95	0.95	0.94	1.06	0.98

Using the same monthly segments, but this time applying a query granularity when ingesting the data, Table 4.4 shows how different querying granularities affect data processing performance. Considering this specific dataset, if the segments are set to the Month level, the query granularity can in turn be set to Day or Week. If Quarter segments are chosen, the query granularity could be set to Day, Week, or Month. For comparison, Table 4.4 includes the characteristics and query processing time obtained with the use of the SMonth table and the corresponding ones when aggregating this table to the Day (SMonth_QDay) or Week (SMonth_QWeek) query granularity. As can be seen, the use of the query granularity has a positive impact on the required storage space. Reducing storage space has, in turn, a positive impact on performance, decreasing the time needed to process the queries. There is, however, one downside to this data organization strategy, namely, the loss of detail for some drill-down interactive analyzes. This happens because the raw data is aggregated and stored in that level of detail.

Table 4.4. Example of different query granularities in Druid. Adapted from Correia et al. (2018).

Table	Number of Segments	Average Size per Segment (MB)	Average Rows per Segment (x 106)	Q2.1 (sec)	Q2.2 (sec)	Q2.3 (sec)
SMonth	80	1060	22.50	4.63	1.57	0.98
SMonth_QDay	80	630	22.33	3.96	1.84	0.91
SMonth_QWeek	80	574	21.70	4.20	1.89	0.88

© FCA

As was already mentioned, besides segments and query granularity, partitions (hashed or dimension based) can also be used. Following the previous example, the effect of applying hashed partitions over the data is shown in Table 4.5. For both query granularities, Day and Week, five and 10 partitions were used, splitting the original segments into five or 10 smaller shards, respectively. Once again, the previously shown scenarios are maintained in the table to facilitate the comparison of the different data organization strategies.

Table 4.5. Example of different number of hashed partitions in Druid. Adapted from Correia et al. (2018).

Table	Number of Segments	Number of Shards	Average Size per Segment/ /Shard (MB)	Average Rows per Segment (x 10^6)	Q2.1 (sec)	Q2.2 (sec)	Q2.3 (sec)
SMonth	80		1060	22.50	4.63	1.57	0.98
SMonth_QDay	80		630	22.33	3.96	1.84	0.91
SMonth_QDay_HP5		400	141	4.47	2.40	1.44	0.77
SMonth_QDay_HP10		800	74	2.23	9.30	1.48	0.87
SMonth_QWeek	80		574	21.70	4.20	1.89	0.88
SMonth_QWeek_HP5		400	124	4.34	2.23	1.30	0.86
SMonth_QWeek_HP10		800	64	2.17	8.95	1.65	0.87

Using hashed partitions increases the storage space needed, when compared to the query granularity scenario without any partitions. When five partitions are defined, query granularities improve the performance, reducing the time needed to process the data. When data is over-partitioned, which happens when 10 hashed partitions are used, the time needed to process the queries usually increases, specifically for the first query of the group, Q2.1. For this query, the time required to process the data when five to 10 hashed partitions are used is around four times higher. Therefore, it is important to balance the distribution of the data. Partitions can be beneficial as long as using them does not result in an excessive number of small files which then need to be processed in order to answer a specific query. Regarding this point, it is important to recall the recommendations available in Druid's documentation, which mentions that segments or shards should have around 5 million records and take up an average of 500MB to 1GB of storage space (Druid, 2018). In the above results, although the best scenarios (SMonth_QDay_HP5 and SMonth_QWeek_HP5) have shards smaller than 500MB, the average number of rows per shard is much closer to the recommended value, suggesting that this indicator seems to be more relevant for query performance than the required storage space.

Given Druid's capacity to enhance OLAP-querying capabilities over large datasets, integrating it with Hive seems advantageous for implementing a BDW. Hive can store the detailed data and support heavy large-scale brute force analytics, while Druid can store materialized views of the data, for processing very specific and detailed queries over a large number of dimensions, increasing the performance of a low latency data analytics system. Druid is usually better than most SQL-on-Hadoop systems such as Hive when it comes to answering specific queries involving finding a few rows within a file with millions of them (Shanklin, 2017).

Druid's performance on low latency analytics is due to the combination of column-oriented storage and inverted indexing. While column-oriented storage minimizes I/O costs, inverted indexes allow drill-down operations at extremely fine levels of detail (Shanklin, 2017). As can be seen in Figure 4-36, specific rows that are needed to answer a query can be loaded using an inverted index.

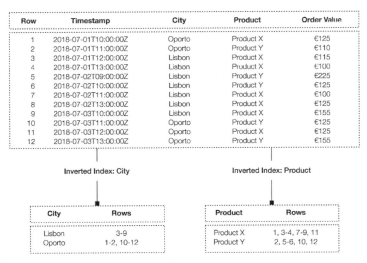

Row	Timestamp	City	Product	Order Value
1	2018-07-01T10:00:00Z	Oporto	Product X	€125
2	2018-07-01T11:00:00Z	Oporto	Product Y	€110
3	2018-07-01T12:00:00Z	Lisbon	Product X	€115
4	2018-07-01T13:00:00Z	Lisbon	Product X	€100
5	2018-07-02T09:00:00Z	Lisbon	Product Y	€225
6	2018-07-02T10:00:00Z	Lisbon	Product Y	€125
7	2018-07-02T11:00:00Z	Lisbon	Product X	€100
8	2018-07-02T13:00:00Z	Lisbon	Product X	€125
9	2018-07-03T10:00:00Z	Lisbon	Product X	€155
10	2018-07-03T11:00:00Z	Oporto	Product Y	€125
11	2018-07-03T12:00:00Z	Oporto	Product X	€125
12	2018-07-03T13:00:00Z	Oporto	Product Y	€155

Inverted Index: City

Inverted Index: Product

City	Rows
Lisbon	3-9
Oporto	1-2, 10-12

Product	Rows
Product X	1, 3-4, 7-9, 11
Product Y	2, 5-6, 10, 12

Figure 4-36. Inverted indexes in Druid.

Druid and Hive can complement each other to support SQL-based analytical capabilities. While Hive provides a standard SQL functionality with HiveQL, Druid's SQL native implementation is very recent and does not aim to support operations such as heavy joins. Druid can join its internal data using small dimension tables loaded from an external system, using a functionality called query-time lookup, but does not support large-scale joins (Shanklin, 2017).

Integrating both technologies, Druid can take advantage of the SQL functionality of Hive, as Druid's data can be queried using Hive or linked to external applications using Hive-based ODBC/JDBC connections, while Hive can take advantage of the historical or real-time data in Druid, stored in a column-oriented store using inverted indexes (Shanklin, 2017). As shown in Figure 4-37, raw data based on a star schema (for instance) can be stored in Hive, whereas the pre-aggregated views, based on the most common queries, can be stored in Druid, taking advantage of the processing capabilities of this tool.

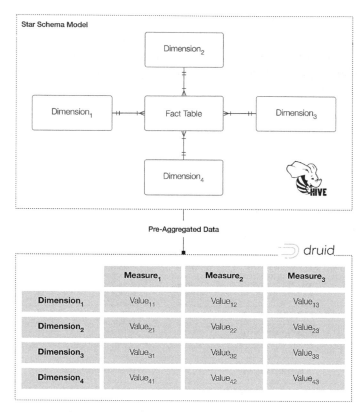

Figure 4-37. Integration of Hive and Druid. Adapted from Shanklin (2017).

To end this chapter, Table 4.6 lists the main features of Hive and Druid, highlighting the strongest access pattern of each technology, namely Hive's large-scale brute-force analytics and Druid's very specific queries with large number of dimensions.

Table 4.6. Features of Hive and Druid. Adapted from Shanklin (2017).

Tool	Strongest Access Pattern	Features
Hive	Large-scale brute-force analytics	Joins Subqueries Windowing functions Transformations Complex aggregations Advanced sorting UDFs
Druid	Very specific queries with large number of dimensions	Dimensional aggregates Top-N queries Min/Max values Time series queries

5 DESIGN AND IMPLEMENTATION OF BIG DATA WAREHOUSES

INTRODUCTION

With the appearance of Big Data techniques and technologies, practitioners started looking for ways of exploring the advantages that Big Data could bring to their businesses. However, the community faced a severe lack of well-defined approaches for building structured analytical repositories capable of storing and processing huge amounts of data, with different formats, and flowing at different velocities. Taking this into consideration, this chapter presents the concept of BDW and, more importantly, a scientifically evaluated approach for designing and implementing these complex systems. The proposed approach contains a set of models and methods to help practitioners in this area. There is a special focus on various aspects: i) the logical components of the BDW; ii) the way data flows throughout these components; iii) the technologies that can be used to implement the various components of a BDWing system; and iv) efficient data modeling to ensure the adequate storage and retrieval of Big Data for advanced decision-making within organizations.

5.1. BIG DATA WAREHOUSING: AN OVERVIEW

The DW concept has a long history, and the need to access, analyze, and present data in appropriate forms to support fact-based decision-making has existed across organizations for a long time (Kimball & Ross, 2013). A DW is a repository that consolidates information about the organization, leveraging a vast range of analyzes developed by several users. Traditionally, it consists in a database that maintains a historical record of the organization, which is periodically extracted from OLTP sources. The DW is designed to access multiple records at a time and it is optimized to support analytical tasks (e.g., predefined or *ad hoc* queries, reports, OLAP, and data mining). OLAP is a common analytical task associated with the DW, mainly consisting in multidimensional structures capable of executing several tasks designed to obtain a desired perspective on the data (Santos & Ramos, 2009). In short, the DW concept is commonly defined as a *"subject-oriented, integrated, non-volatile, and time-variant collection of data in support of management's decisions"* (Inmon & Linstedt, 2014, p. 315), as well as a *"single version of the truth"* (Kimball & Ross, 2013, p. 407).

Over the last decades, traditional DWs have been recognized as an enterprise data asset, but the evolution of advanced analytics (e.g., data mining, statistics, and complex queries), the growth of data volume, and real-time needs to analyze fresh data are driving changes in DW architectures (Russom, 2014). Nowadays, the DW is evolving, it is being extended and modernized to support advancements in technologies and business requirements, to prove its relevance in the era of Big Data. DW modernization is a top priority for professionals; and surveys show that DWs are evolving dramatically and there are several opportunities to improve and modernize them, since organizations view them as key assets in today's businesses (e.g., analytics, data-driven decision-making, operational efficiency, and competitive advantages) (Russom, 2016).

In this modernization process, however, some challenges arise, such as inadequate data governance, lack of skills, cost of implementing new technologies, and difficulties in conceiving a modern solution that can ingest and process the ever-increasing amounts or types of data. According to Russom (2016), the average DW stores between 1TB and 3TB of data and it was predicted that it would store between 10TB and 100TB by 2018. Organizations need to consider the modernization of their DW architectures whenever some of the following questions emerge (Chowdhury, 2014):

- Is the current platform limited by the amount of data that needs to be processed?
- Is the DW a useful repository for all the data that is generated and acquired? Or is some data being left unprocessed due to current restrictions?
- Do we want to analyze non-operational data and use new types of analytics?
- Do we need to ingest data quicker?
- Do we need to lower the overall cost of analytics?

Research related to the BDW is mainly divided into five topics: the characteristics and design changes of DWs in Big Data environments; DWs on NoSQL databases; storage technologies, optimizations, and benchmarking for BDWs; advancements in OLAP, query, and integration mechanisms for BDWs; and implementations within specific contexts.

Some of the main features that define a BDW are:

- Parallel/distributed storage and processing of large amounts of data;
- Scalability (capacity to accommodate more data, users, and analyzes);
- Elasticity to provide a more efficient way of scaling-out and scaling-in depending on the organizational needs;
- Flexible storage, including semi-structured and unstructured data;
- Real-time capabilities (stream processing, low latency, and high frequency updates);
- High performance with near real-time response;
- Mixed and complex analytics (e.g., *ad hoc* or exploratory analysis, data mining, text mining, statistics, machine learning, reporting, visualization, advanced simulations, and materialized views);
- Interoperability in a federation of multiple technologies;
- Fault-tolerance, mainly achieved through data partitioning and replication;
- The use of commodity hardware to reduce the costs of implementation, maintenance, and scalability.

Hadoop and NoSQL databases are typically mentioned either as a replacement for traditional DW or as a mean to augment its capabilities (e.g., ETL, data staging, and preprocessing of unstructured data), thus forming a federation of different technologies that provide the aforementioned characteristics. Figure 5-1 presents a conceptual model of the BDW, which illustrates the features previously discussed and the strategies discussed in the following text.

Designing a BDW should focus both on the physical layer (technological infrastructure) and the logical layer (data models, data flows, and interoperability between components). Augmenting the capabilities of traditional DWs with new technology is a valid approach and, arguably, the one currently preferred by most organizations; this strategy is known as "lift and shift". However, "rip and replace" strategies wherein traditional DWs are fully replaced due to their limitations in Big Data environments will become more common (Russom, 2014, 2016). The "lift and shift" strategy creates a federation of different technologies and may represent a change of perspective from a data-driven view of the DW to a use case driven view (Clegg, 2015). Whereas data modeling was previously the main concern, it is now replaced with finding the right technology to meet emerging demands, and this can increase the risk of ending up with uncoordinated data silos.

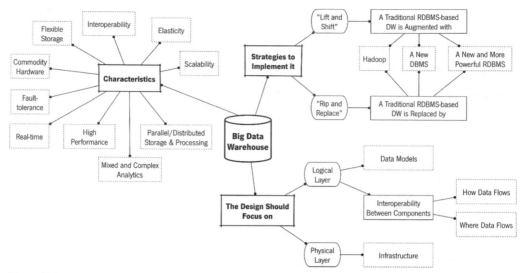

Figure 5-1. A conceptual model of the BDW.

Although current research has significant value, it only contributes to improve specific characteristics of a BDW by promoting some existent technology, proposing a new one, or developing a specific implementation for a particular use case. This phenomenon among research on Big Data is conspicuous with most of the contributions focusing on "selling" their technique or technology. As a result, there is little prescriptive research on BDW, since there is no integrated approach for designing BDWs, such as those developed for traditional DWs, like the well-known approaches from Kimball or Inmon. This gap can be mainly attributed to a change of perspective: a shift from a data-driven view of the DW to a use case driven view, and also due to the fact that Big Data is still a relatively young research topic. However, as Clegg (2015) claims, it would be a mistake to discard decades of architectural best practices based on the assumption that storage for Big Data is not relational nor driven by data modeling principles or guidelines.

The fact that there is a significant number of works related to SQL-on-Hadoop proves that the data structures that have been implemented for many years are more relevant than ever, although they need to be modified and optimized for Big Data contexts. Of course, there is unstructured data that does not adequately fit into these data structures, but there are also techniques to extract value from it, and then use it to fuel a BDW (e.g., data mining, text mining, and machine learning). The problem, however, is rather the significant gap between "this is what a BDW should be" and "this is how one designs and implements it", which then leads to a use case driven approach, which is primarily concerned with choosing the right technology to meet specific demands. Since approaches are normally use case driven, the knowledge and guidelines that can be retrieved from one implementation and applied to another one are only possible because the circumstances are more or less the same. As a result, there is little gradual and iterative knowledge being created, which is crucial for fundamental advancements in this area.

Among the scientific contributions related to this topic, there are already some best practices and general guidelines, but they do not focus on both the physical layer (technological infrastructure) and the logical layer (data models, data flows, and interoperability between components) to implement the characteristics of a BDW, with adequate and detailed demonstration, discussion, and evaluation. Doing so would have a major impact on the scientific and technical community related to BDWing, since it would lead to a contribution in which models (representation of data structures and components), methods (structured practices) and instantiations (prototypes and implemented systems) are tightly coupled. Such approach can lead to a prescriptive contribution for designing and implementing BDWs according to their features of parallel/distributed storage and processing, scalability, elasticity, real-time, high performance, mixed and complex analytics, flexible storage, interoperability, fault-tolerance, and commodity hardware.

The approach proposed in this book is prescriptive: it aims to help researchers and practitioners to build BDWs and provides background knowledge for future research. It significantly extends the current research on BDWing, which is scarce and scattered, by including prescriptive models and methods that can be used as a guide for designing and implementing these complex analytical systems. The approach is based on the "rip and replace" strategy (Russom, 2016), which involves discarding traditional RDBMS-based DWs and replacing them with state-of-the-art Big Data techniques and technologies. It is an approach that aims to address the characteristics of a BDW (see Figure 5-1), focusing on both the logical and physical layers. This chapter describes the proposed approach, presenting its prescriptive models and methods, namely: a model of logical components and data flows; a method for data CPE processes; a model for the technological infrastructure; and a method for data modeling focusing on data storage and analytics.

5.2. MODEL OF LOGICAL COMPONENTS AND DATA FLOWS

The logical components included in the proposed approach (Figure 5-2) are defined according to the components present in the NBDRA (NBD-PWG, 2015), since our approach complies with current standards and trends set by the Big Data community. Obviously, our approach presents some significant modifications and also extends the NBDRA with new components, since the latter is a general architecture for Big Data solutions and, therefore, not specifically designed towards BDWs like our proposed approach. The approach hereby proposed also takes into consideration relevant guidelines provided by previous published works, such as the Big Data Processing Flow by Krishnan (2013) and the Data Highway Concept by Kimball and Ross (2013). Furthermore, our approach encourages compliance with three of the main principles of the Lambda Architecture (Marz & Warren,

2015): first, one should store data at the highest level of detail (e.g., raw data in the distributed file system component), since it may serve future analytical purposes that have not yet been planned; second, whenever possible, one should model data structures to store a set of immutable events, avoiding updates to existing data, in order to simplify the BDWing system; finally, data at different speeds certainly has different requirements and, therefore, different logical components for batch and streaming data must be taken into consideration. The model of logical components and data flows (Figure 5-2) is divided into six main components: data provider; data consumer; Big Data application provider; Big Data framework provider; system orchestrator; and security, privacy, and management.

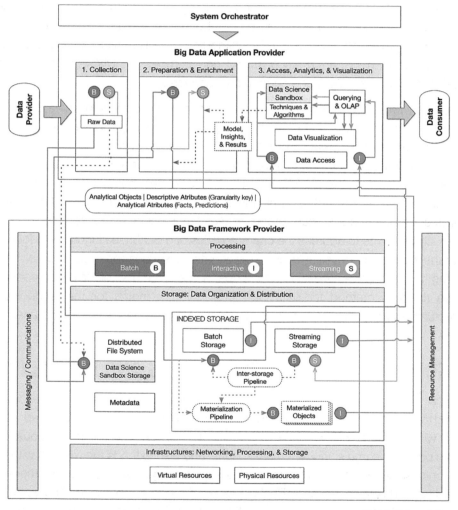

Figure 5-2. Model of logical components and data flows.
Dashed components are seen as optional depending on the implementation goals.

5.2.1. Data Provider and Data Consumer

The data provider component represents each actor introducing new data into the BDW (e.g., every person, sensor, computer, smartphone, and Web stream). Therefore, this component represents several data sources, external or internal, online or offline, collected automatically or manually. Some of the main responsibilities of the data provider are the following: ensure adequate data privacy and security; enforce access rights; make data available through suitable interfaces; and provide adequate metadata. In contrast, a data consumer represents an end-user or an external system that can perform the following actions: search and download data; analyze data (e.g., execute *ad hoc* queries, and train/ test data science models); construct or consume reports and other data visualization mechanisms (e.g., dashboards); and include data in business processes. To access the data available in the implemented BDWing system and protected by the security, privacy, and management component (e.g., authentication and authorization mechanisms), the data consumer is able to use the interfaces made available by the Big Data application provider through demand-based interaction, where the data consumer initiates the interaction and waits for a response (NBD-PWG, 2015).

5.2.2. Big Data Application Provider

The Big Data application provider component is responsible for ensuring three relevant stages undergone by the data flowing throughout the different BDW components:

1. **Collection** – in this stage, the data is collected from data providers, arriving at the BDWing system in the rawest state possible. The data can arrive at the system at two different velocities, batch or streaming. Data arriving in batches is immediately stored in a distributed file system, since one of the main challenges in Big Data is variety (different structures, types, and sources), and this file system is a component that allows the storage of any variety and volume of data. This data will be processed in the next stage. In contrast, if the data is arriving in a streaming fashion, it will not need to be stored yet, it will flow to the preparation and enrichment stage. However, an alternative route in the streaming flow can exist, storing streaming raw data in the distributed file system, as Figure 5-2 shows, making this data available for further tasks, such as disaster recovery or to train data science models on unmodified streaming data.

2. **Preparation and enrichment** – the batch data previously stored is extracted from the distributed file system to prepare it and enrich it to provide analytical value. The same happens with the streaming data, although it arrives at this stage directly from the collection stage, as previously explained. The preparation and enrichment of data can include all sorts of cleansing, integration, transformation, and aggregation processes. New attributes can also be created and derived from the raw data, without any limitation,

© FCA

as well as the extraction of hidden patterns in unstructured data (e.g., image, video, and text). These processes cannot only take as input the new data arriving at the system, but also read the data already stored in the BDW, establishing comparisons and trends, for example. Besides finding patterns in unstructured data using data science techniques (e.g., text mining), these processes can also include predictions from previously trained data science models based on problems such as classification, regression, clustering, and time series forecasting. It must be remembered that the goal at this stage is to prepare and enrich data to serve the current business goals and expectations, whether they are based on facts or on predictions made by previously trained data science models (this data flow is surrounded by a dashed circle in Figure 5-3).

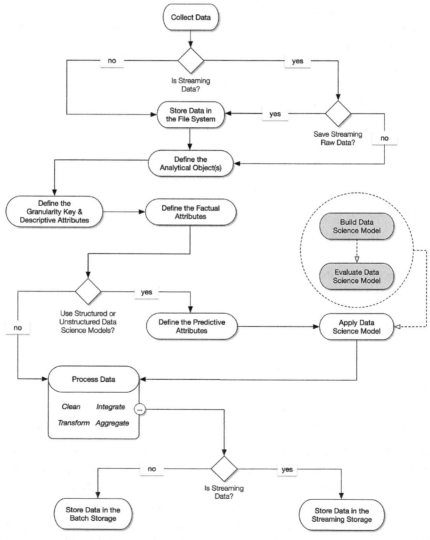

Figure 5-3. Method for CPE processes.

The way to implement these processes differs according to the velocity of the data (batch or streaming), since different velocities typically require different paradigms and technologies, but the essential steps are similar. In the proposed approach, batch and streaming data follow different routes, but are prepared and enriched with the same goal in mind: fuel the analytical objects to which the data belongs, structuring data according to their granularity key, descriptive attributes, and analytical attributes, whether they are facts or predictions extracted from the raw data being collected. The method for data modeling detailed in section 5.4 explains these concepts, detailing how data should be modeled in the BDW. After this stage, the data is stored in its corresponding indexed storage component: batch data is stored in the batch storage; and streaming data is stored in the streaming storage. Figure 5-3 illustrates the proposed method for CPE processes, summarizing the steps described above.

3. **Access, analytics, and visualization** – this is the third and final stage of the data in a BDWing system. In this stage, the data consumers have access to the data in two fundamental ways: batch and interactive access. A batch access can be used for complex visualization tasks (e.g., deep and complex reporting) or for training and testing data science models, which typically require intensive computation. The data science sandbox (see Figure 5-2) is a crucial component of the BDW, where data scientists can explore the data stored in the distributed file system and indexed storage components, create models, extract useful insights, and use the results to improve or create new analytical objects. Therefore, preparation and enrichment processes can also begin with batch jobs originated in the data science sandbox, without the need for a collection stage, basically meaning that the data science sandbox is just considered as another data provider in this context. These are tasks that do not necessarily require an interactive response time and, therefore, can be seen as batch-oriented. In contrast, tasks such as *ad hoc* querying, OLAP, and exploratory data visualization require an interactive behavior to keep the data consumer engaged in the current analysis and exploration of data. The tasks enumerated previously are relatively similar to the ones performed in a traditional DW (e.g., reporting, data visualization, and exporting data to run data mining algorithms). Obviously, there are some particularities to take into consideration in Big Data environments (volume, variety, and velocity), but there is no need to propose a specific method to perform these tasks, since Figure 5-2 is self-explanatory.

5.2.3. Big Data Framework Provider

This logical component includes the subcomponents related to the resources that are necessary to provide an adequate distributed storage and processing platform for the BDW, as well as an adequate communication with external actors (e.g., data providers and data consumers). Therefore, this component is mainly related to infrastructural concepts, such as messaging/communications, resource management, infrastructures (physical or virtual), data processing paradigms (batch, interactive, and streaming), and data organization and distribution paradigms (distributed file system and indexed storage).

5.2.3.1. Messaging/Communications, Resource Management, and Infrastructures

The messaging/communications component responds to the need of ensuring reliable queuing and data transmission between nodes in a cluster that scales horizontally (NBD-PWG, 2015). An important point to remember is that scalability is one of the main characteristics of BDWs. Messaging/communications techniques must also ensure adequate fault-tolerance when nodes fail. In the design and implementation of BDWs, this component is relatively transparent to the stakeholders involved in the project, since it is directly related to the technologies chosen to implement the logical components of the BDWs (see section 5.3). Each technology may implement different messaging/communications techniques, and, depending on the application context, one may prefer a specific technique that better meets the requirements of the project. This is something that stakeholders should be aware of, but it is often transparent to the team installing and configuring these technologies.

Regarding the resource management component, it represents a relevant concern with the technologies that ensure distributed storage and processing, which are an adequate way of achieving scalability in a BDW. These technologies must efficiently manage the resources available in the cluster, namely CPU and memory. Inadequate management of resources may severely impact the performance of the BDW. One of the main concepts to bear in mind regarding this component is "data locality", since data is too big to be transferred from the storage nodes to the processing nodes through the network (NBD-PWG, 2015). Therefore, the processing needs to be closer to the storage (C. L. P. Chen & Zhang, 2014), this is typically achieved co-locating the processing and storage nodes in the cluster. Like the messaging/communications component, the resource management component is directly related to the technology chosen to implement a specific logical component. Each technology may use different techniques to manage resources.

Finally, the logical component related to the infrastructures highlights all the physical and/or virtual elements necessary to run the tasks assigned to each component in the BDW, including the network for transfering data, the CPU and memory for providing adequate data processing, and the storage for providing data persistence. Physical resources represent the hardware used across the different nodes in a horizontally scalable cluster. In contrast, virtual resources are frequently used to achieve an elastic and flexible allocation of physical resources, typically referred to as IaaS. Although Big Data technologies can be deployed on virtualized environments, the majority of them are designed to run directly on physical commodity resources (NBD-PWG, 2015), providing efficient I/O by distributing multiple CPUs, memory units and disks across a cluster of commodity machines based on a shared-nothing architecture.

5.2.3.2. Processing

In a BDW built using the proposed approach, there are three types of processing according to the different levels of latency, namely batch, interactive, and stream processing. Generally, the boundaries between these three types of processing are not clear. However, in this book, they are defined as follows:

- **Batch processing** – this type of processing involves latencies ranging between several minutes and hours. Examples of batch processing may include: the periodic CPE of vast amounts of historical data from data providers; the processing of deep and complex reports or *ad hoc* queries; and the training of data science models, which involves complex and processing-intensive data mining and text mining algorithms. These tasks are ideal for running in the background without the need for user intervention;

- **Interactive processing** – this type of processing is used to provide query execution times ranging from milliseconds to a few tens of seconds, depending on the infrastructure and data volume. There are a few data organization, distribution, and modeling strategies used in this work that allow for this level of latency even with ever-increasing amounts of data. Such strategies include: data denormalization; data partitioning; and inter-storage and materialization pipelines (see subsection 5.2.3.3 and section 5.4);

- **Stream processing** – in this work, this type of processing only concerns the latency in data CPE processes, meaning that data consumers do not have direct access to the data being streamed. Instead, streaming data is stored in the streaming storage component and it becomes immediately available to the data consumers through interactive processing supported by this data storage component. Regarding the levels of latency, streaming data should arrive at the streaming storage within milliseconds or a few seconds, in order to be immediately available to data consumers. However, when preparing and enriching streaming data, sometimes it is useful to create micro batch jobs to perform specific operations, such as small aggregations, window operations, application of data science models, and merging streaming data with batch data to establish trends. Micro batches can be seen as having a significantly small batch of data records, instead of handling them individually. The size of a micro batch job is often customizable in several streaming technologies. When this type of operations is needed, the data arrives into the BDWing system in a streaming fashion, but may only

© FCA

be available within a few tens of seconds or even minutes, depending on the size of the micro batch. Micro batches can also help improving the throughput of the data flow, only requesting an insert operation on the streaming storage component when a micro batch is completed, instead of creating a request for each data record.

5.2.3.3. Storage: Data Organization and Distribution

Data organization and distribution is a crucial aspect in the proposed approach. It is designed to provide a flexible and scalable data storage solution that is aware of data volume, variety, and velocity, without discarding a data modeling method. In this context, the storage design philosophy presented here is based on two relevant components: the distributed file system, which is an unstructured data storage solution, wherein data does not necessarily need to have a specific schema nor does it need to be modeled in a specific way; and the indexed storage component, wherein the data must comply with specific structures, although they are based on a flexible modeling technique suitable for BDWs, as detailed in section 5.4. These two components are also related to the logical component that provides all the metadata for the data stored in the file system and in the indexed storage components (e.g., file locations, data types, descriptions, and relevant timestamps).

Distributed File System

Making an analogy with the traditional DWs, the distributed file system can be seen as an empowered staging area, wherein raw data is not only stored for later preparation and enrichment, but also for training data science models based on structured or unstructured data, since a file system adequately supports schema-less data sources. Therefore, data scientists can use this file system as a sandbox to explore the data and discover hidden patterns, providing useful insights to support the decision-making process. Taking a closer look at Figure 5-2, it can be observed that data scientists can also use this component to store the results of queries submitted to the indexed storage component, and use these results to create or improve data science models based on several techniques and algorithms. These models and insights can then be included in further data CPE processes, combining them with data arriving at the BDW and storing the result in the analytical objects (see section 5.4) stored within the indexed storage component.

This approach provides adequate flexibility to freely explore the data in its raw state, to combine it with previously stored data, if applicable, and to make sure that the sandbox findings flow to the indexed storage component, which ensures that the analytical requests from data consumers are fulfilled. Take as an example a company that sells jewelry in several countries. This company collects data from its point of sale systems using batch processing, as well as unstructured text from social media using stream processing. Data scientists use the unstructured data stored in the distributed file system to train a text mining model for extracting sentiments about the different types of jewelry in several countries. In the indexed storage, the company stores an analytical object for the sales data and an analytical object for the sentiments expressed about the different types of jewelry in each country. Meanwhile, after days of querying the data stored in the indexed storage component and saving the findings in the distributed file system, a certain data scientist can begin classifying a sale as "expected" or "unexpected", which results from the comparison between the jewelry being sold and the sentiments expressed for that product in the country in which the sale is being made. This is an example of the usefulness and flexibility of the distributed file system in a BDWing system built using the proposed approach, to complement or create analytical objects stored in the indexed storage component.

Indexed Storage

In contrast to the distributed file system, the indexed storage is a component intended for data modeling, i.e., data needs to be structured according to a specific data model. However, in this work, the data model based on analytical objects offers significant flexibility, while maintaining a structured schema suitable for querying and OLAP. This modeling method is further discussed in section 5.4. In this subsection, the focus is on the logical components responsible for storing these analytical objects. The indexed storage component is divided into two main storage types, namely the batch storage and the streaming storage. Nevertheless, the data modeling approach is the same for the two types of storage, storing all the data in analytical objects and their descriptive and analytical attributes.

The batch storage component represents a repository of analytical objects that are refreshed less frequently, since the data only arrives in a batch-oriented fashion and, therefore, the time interval between updates is usually of several minutes, hours, days, weeks or months, for example. In contrast, the streaming storage component stores analytical objects that are refreshed frequently, since the data arrives through streaming mechanisms and, therefore, updates usually happen within time intervals of milliseconds, seconds, or a few minutes (for large micro batches). A relevant component related to these two storage types is the inter-storage pipeline, which is responsible for transferring data between the streaming storage and the batch storage. Consequently, the same analytical object may exist simultaneously in these two storage components. This may happen if the technology being used to support the streaming storage has fast random access to data, but it is not optimized for fast sequential access. In contrast, if the technology is the same for both storage types, so either balanced or more optimized for fast sequential access and not for fast random access, frequently, the inter-storage pipeline may only need to

© FCA

execute background jobs to distribute data in a more efficient way, such as, for example, merging many small files originated by the streaming process into one larger file, since small files can become a problem in Hadoop (Mackey, Sehrish, & Wang, 2009). An important point to remember is that, internally, indexed storage systems also persist data as files. The inter-storage pipeline is optional, depending on the infrastructure being deployed to support the BDW, since technology is constantly evolving, and there is increasing interest in exploring storage systems that adequately support both fast sequential access and fast random access in Big Data environments, as discussed in section 5.3.

Other optional subcomponents included in the indexed storage component are the materialization pipeline and the materialized objects. A materialized object is an object that stores the results of a query executed over one or more analytical objects. The materialization pipeline is the logical component that ensures this materialization process. Materialization can be significantly helpful for improving execution times in contexts where the data consumer consistently submits similar queries to the BDWing system. Moreover, materialization also helps storing the results of deep and complex requests like long-running reports, which would otherwise take a significant amount of time to complete. Materialization may typically impose a trade-off between data timeliness and response times, but there are several contexts where the data consumers do not need the most recent data available in the BDW. Nevertheless, materialized objects can be refreshed when a new batch of data arrives into the system or when the inter-storage pipeline runs a background job (see Figure 5-2). Consequently, the materialization pipeline can either re-process the whole materialized object, or perform an incremental change by reading it and complementing it with new data.

To conclude this subsection regarding the indexed storage component, it is important to highlight that the analytical objects stored either in the batch storage or in the streaming storage can be organized and distributed using two relevant concepts: partitioning and bucketing/clustering (Thusoo et al., 2010). These two concepts can have a strong influence over query performance in certain contexts. When relying on an indexed storage that uses partitioning, all the data belonging to an analytical object is stored as many small pieces of data inside the storage system, dividing a large dataset into many small and more manageable parts that can be accessed individually, without the need to search the entire dataset. An example of partitioning involves storing an analytical object such as sales transactions using a separate storage location for each year, month and/or day. If one needs to analyze the sales of the previous month, the indexed storage system only has to scan the partition corresponding to that month. Partitioning can be significantly helpful when data consumers have a well standardized access to data, such as querying the data stored in analytical objects for a certain period (e.g., year, month, and day) or place (e.g., country, region, and city). Consequently, partitioning improves the performance of queries when the typical filtering attributes are used to partition the dataset. Partitioning can also be significantly useful when data is loaded in periodic batches or in batches corresponding to certain places, since a partition can be assigned for each batch.

Bucketing/clustering is a technique that ensures that a range of records are stored in the same group/bucket or sorted in a certain way, according to the attribute(s) used for bucketing/clustering. The way bucketing is physically implemented differs according to the technology, i.e., some storage technologies may group a range of values in the same file, whereas others can organize the values and make sure they are stored in a sorted fashion. Following the example of sales transactions, if a certain organization has several sales employees, using a bucketing/clustering technique with the identification of the employee, the indexed storage can store the transactions of the same employee in the same bucket, or make sure the transactions are sorted according to the identification of the employee. Partitioning and bucketing/clustering can be used together, and query performance can be significantly impacted when adequate strategies are taken into consideration (E. Costa et al., 2018).

5.2.4. System Orchestrator and Security, Privacy, and Management

The system orchestrator has an overarching role, involving various actors (humans and/or software) that manage and orchestrate the daily operations of the BDWing system. The system orchestrator aims to configure and manage other components of the architecture, to sustain the workloads that are being constantly executed. Its tasks include: assigning/ provisioning the Big Data Framework Provider (see subsection 5.2.3) to physical or virtual nodes; provide Graphic User Interfaces (GUIs) for specifying and managing workloads; and monitor the system and its workloads through the security, privacy, and management component, taking into account requirements and constraints such as business requirements, policies, architectural design choices, and resources (NBD-PWG, 2015).

In the proposed approach, the security, privacy, and management component is an overarching aspect that is related to all other components in the BDWing system. Managing such complex system typically involves several considerations at a massive scale, while the system performs multiple tasks in a production cluster with various nodes. Among the tasks concerning this component, the following can be highlighted (NBD-PWG, 2015):

- Policy, metadata, and access management (authentication and authorization);

- Provide adequate encryption capabilities at networking or storage levels (if needed);

- Provide adequate auditing capabilities;

- Disaster recovery in case of data loss;

© FCA

- Provide adequate monitoring mechanisms for the resources and performance of the system;

- Make adequate platforms available for resource allocation and provisioning, as automated as possible;

- Configure and manage the installed software.

5.3. MODEL OF TECHNOLOGICAL INFRASTRUCTURE

The model of logical components and data flows is an artifact for the design of BDWing systems, whereas the technological infrastructure model is an artifact to implement the former. It focuses on the technologies that can instantiate each logical component, as well as on the hardware that can be used to deploy the BDWing system. In this section, Figure 5-4 presents the technological infrastructure model, including several state-of-the-art technologies for each logical component of the BDW presented in Figure 5-2. Therefore, a direct association can be made between the two figures, providing a coherent view and simplifying the design and implementation phases of BDWing initiatives. The colors (blue, orange, and green) correspond to the types of processing depicted in Figure 5-2 (batch, interactive, and streaming, respectively). The technologies presented in Figure 5-4 are merely suggestions, which are based on several Hadoop-related projects, and should not be seen as preferences over any other technology that researchers and practitioners may find better suited for implementation. This is why the model explicitly shows there is space for other possibilities. For each logical component, several suitable technologies are presented, which must be seen as alternatives or complements, not as mandatory implementations. Finally, Figure 5-4 also shows how these technologies are supported by a scale-out infrastructure, deployed on-premises or in the cloud, either using physical or virtual resources.

Starting with data collection, Flume and Kafka are suitable technologies that can be used to collect data in a streaming fashion. In contrast, Sqoop can be used to move batches of data from relational databases into HDFS. There are also ETL tools oriented towards Big Data contexts (e.g., Talend Big Data), which include components for both batch and streaming data collection. However, frequently, these tools, in their open source versions at least, only provide an integrated GUI for submitting tasks to systems such as Flume, Kafka, and Sqoop. Therefore, the technologies mentioned above still have to be deployed on the infrastructure. Furthermore, for specific data collection scenarios, one may need to implement custom collectors developed using well-known programming languages, such as Java, Python, and Scala, either for batch or streaming scenarios.

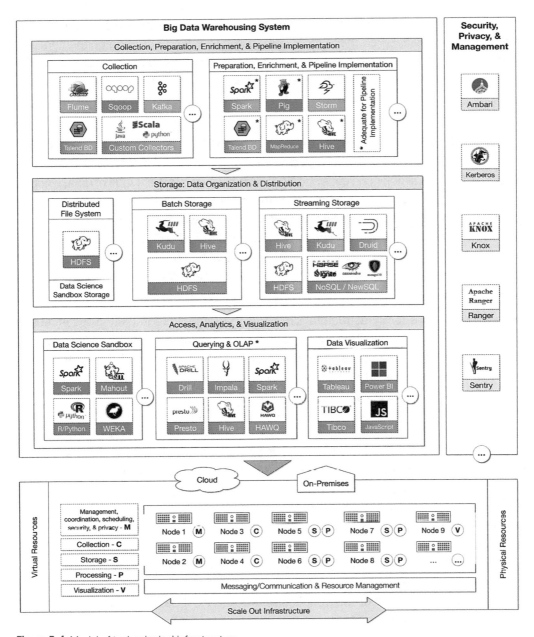

Figure 5-4. Model of technological infrastructure.

Hadoop-related projects like Pig, Hive, and Spark are adequate technologies for data preparation and enrichment using batch processing. Native MapReduce code, although complex, can also be used for that purpose, as well as Talend Big Data. Storm, Spark Streaming, and Talend Big Data can be used for preparation and enrichment via streaming. Nevertheless, as previously mentioned, Talend's open source version typically relies on other components to ensure adequate distributed processing, since its native components may not be scalable. Since these tools comprise a vast set of storage connectors and data processing components, some of them are also adequate for supporting the implementation of the inter-storage and materialization pipelines (namely, the technologies marked with an asterisk in Figure 5-4). The chosen technology will depend on the choice of the storage technologies.

Storage technologies are a crucial aspect of the BDWing system, and maybe one of the most difficult to understand. Regarding Big Data technologies for the distributed file system, HDFS is a simpler choice, since it provides a way to store all kinds of data, whether structured or not. The choice depends on the indexed storage component, i.e., on the batch storage and on the streaming storage. There are two main approaches to choose from: the first one is based on infrastructural simplicity, which alleviates management burden for system orchestrators; the second one is a hybrid approach, which may lead to more efficient refresh processes, but can also bring more challenges for managing the infrastructure.

Assuming one aims for infrastructural simplicity, the same storage technology is reused as many times as possible. Therefore, since HDFS is used as the distributed file system, it can also be used for historical and streaming storage. Hive relies on HDFS to store data, so, technically, using Hive tables to store the analytical objects is as complex as using raw HDFS files. Consequently, using HDFS with file formats intended for analytics such as Parquet and ORC (Huai et al., 2014), or using Hive tables stored in these formats, is the approach with maximum infrastructural simplicity. However, this approach may sacrifice data refresh rates, since streaming mechanisms will have to group data records in larger micro batches, to avoid creating multiple small files, which can cause problems for Hadoop (e.g., a larger metadata footprint in RAM and unsatisfactory NameNode performance) (Mackey et al., 2009), as was briefly mentioned in section 5.2. This phenomenon occurs because HDFS and Hive are currently oriented towards fast sequential access instead of fast random access (more details related to streaming scenarios are provided in section 8.3). The problem is that increasing the micro batch size also increases the interval between data collection and its availability for querying in the BDWing system. Nonetheless, there are many streaming contexts where it is less problematic if data is available only after a few minutes after being collected.

To achieve shorter time intervals between the collection of data and its availability for querying, one can use storage systems oriented towards fast random access, such as NoSQL databases. Another advantage of these systems is their ability to perform random reads or updates on data, which can be useful for certain BDW applications. For instance, these systems enable efficient update operations on records, in cases where it is not feasible to model data as a set of immutable events. However, since these databases are mainly used for OLTP-based workloads (Cattell, 2011), they typically do not perform as well as the fast sequential access systems for OLAP-based workloads. Consequently, choosing NoSQL databases solves the small files problem in Hadoop, but may also bring more infrastructural complexity and slower query execution times for OLAP-based workloads (see results discussed in section 8.3).

Among NoSQL databases, one can also highlight the relevance and possible use of in-memory NoSQL databases like Redis (2018b), or even NewSQL databases like Apache Ignite (2018), if the chosen querying and OLAP system supports these technologies. Some of these technologies may provide faster query execution times, as they sometimes have more optimized in-memory architectures. Again, the modularity of the approach permits flexible implementation choices without changing any significant architectural construct or data modeling guideline. Another adequate technology for streaming scenarios is Druid (Yang et al., 2014), a columnar store that can be used to support interactive and concurrency-heavy applications focusing on slicing-and-dicing, drilling down, and aggregating event data. Druid achieves this by aggregating and indexing time-based data as soon as it arrives to the system, providing sub-second queries over vast amounts of streaming data (Correia et al., 2018). Another adequate use case for Druid is the storage of materialized objects due to its on-the-fly aggregation mechanisms. Although Druid can be used for the batch storage component as well (Correia et al., 2018), this book focuses on its use for streaming and materialization scenarios containing aggregated data indexed by temporal attributes. Like many other Big Data technologies, Druid has limitations (e.g., lack of support for random access operations), and users should conduct a preliminary analysis when choosing storage technologies, because the ecosystem is evolving rapidly.

There are other technologies offering a middle ground between fast sequential access and fast random access, such as Kudu, which supports both scenarios without requiring different storage systems (Lipcon et al., 2015). Kudu can be co-located with other components from the Hadoop ecosystem and, therefore, can be used along with HDFS. Using the same storage system for both batch data and streaming data can also reduce infrastructural complexity, although HDFS should continue to be used as the distributed file system. Furthermore, as previously noted, technology is evolving rapidly, and with the community advancing Hive transactions and streaming support (Apache Hive, 2018), streaming scenarios and update operations in Hive are becoming more streamlined. Currently, when considering implementing BDWing systems, practitioners should spend some time studying how these systems work, as well as their advantages and disadvantages for implementing an adequate and stable storage system for the BDW, since there is no optimal solution for all implementation contexts.

Regarding data access, analytics, and visualization, there are several technologies that can be used for specific tasks. Spark MLlib and Mahout are two machine learning and data mining libraries that rely on distributed processing to extract patterns from a large volume of data. R, Python, and WEKA, for example, can also be used for this purpose, but one should be aware of their limitations in Big Data environments, as was previously discussed in subsection 2.5.3. However, during the last years, these technologies began to include processing components that can establish connections to distributed systems such as Spark and Hadoop. Technically, any machine learning and data mining technology able to process large amounts of data and with adequate connectors to Hadoop-related systems can be used in a BDW data science sandbox. In these terms, technologies such as Tableau, Microsoft Power BI, or TIBCO Spotfire can be used to visualize data. Customized visualizations can be created with custom-made JavaScript applications (e.g., intensive geospatial analytics – see the SusCity data visualization platform in section 9.4). The data visualization tool being implemented needs to provide adequate connectors for querying and OLAP technologies. However, in certain scenarios wherein direct access is required, bypassing the querying and OLAP engine is acceptable. This could be done using native storage drivers (e.g., HDFS, Hive, NoSQL/NewSQL, Kudu, and Druid), or by developing custom-made Web services (e.g., REST Web services), to prevent some incompatibilities, or to ensure higher concurrency and efficiency for certain scenarios demanded by data consumers (e.g., concurrent custom-made Web data visualizations).

The querying and OLAP systems are crucial for BDWing, since they provide an interactive SQL interface to query the data stored in the batch storage and the streaming storage. These systems are frequently mentioned as SQL-on-Hadoop systems, although they also support other data sources like NoSQL databases. There are several alternatives, such as the following: Hive (on Tez) (Huai et al., 2014); Drill (Hausenblas & Nadeau, 2013); Impala (Kornacker et al., 2015); Spark SQL (Armbrust et al., 2015); Presto (2018); and HAWQ (L. Chang et al., 2014). Benchmarking different SQL-on-Hadoop systems is advisable when implementing a BDWing system (Rodrigues et al., 2018; Santos et al., 2017), since it helps evaluating whether response times, scalability, and SQL compatibility meet the established requirements. Furthermore, evaluating their connectivity with the storage and data visualization systems is crucial in order to implement an adequate and interoperable BDWing system.

Such complex technological infrastructure needs to be secured and properly managed, ensuring compliance with security and privacy policies, as well as providing a set of mechanisms to monitor the behavior of the infrastructure and respond to threats accordingly, if necessary. In this context, Ambari can be used to provision, manage, and monitor a Hadoop cluster supporting the BDWing system. Regarding security, there are several technologies that can be used depending on the specific requirements: Kerberos can provide secure authentication for users and resources; Knox can provide perimeter security, hiding the details of the cluster's access points and blocking services; Sentry can be used to define adequate authorization policies to access data; and Ranger, which is similar to Sentry, provides a centralized platform for policy administration, authorization, auditing, and data protection (HDFS encryption). There are other ways of ensuring data security and privacy, such as using specific encryption mechanisms or access control lists provided by different technologies.

To conclude this section, there are some relevant guidelines that should be taken into consideration when deploying an adequate infrastructure for BDWing:

1. Plan the infrastructure focusing on horizontal scalability scale out to reduce costs and leverage the full potential of emergent Big Data technologies like Hadoop. *"Because Hadoop uses industry-standard hardware, the cost per Terabyte of storage is, on average, ten times cheaper than a traditional relational DW"* (Krishnan, 2013, p. 273).

2. Co-locate storage and processing nodes in the cluster to avoid moving data between nodes, causing bottlenecks in the network (C. L. P. Chen & Zhang, 2014). As can be seen in Figure 5-4, storage and processing nodes are always co-located. This means that querying and OLAP technologies should be installed in all the storage nodes, thus data is not moved across nodes when data consumers submit a request.

3. Implement a Just a Bunch of Disks (JBOD) configuration for each storage node. If a Redundant Array of Independent Disks (RAID) configuration must be used, implement a RAID-0 strategy (Xu, Luo, & Woodward, 2012).

4. Implement at least a 1-gigabit Ethernet network infrastructure (Shvachko, Kuang, Radia, & Chansler, 2010).

5.4. METHOD FOR DATA MODELING

This section presents the data modeling method for designing the data structures stored in the indexed storage component of the BDW. It discusses how data should be modeled according to specific data structures denominated analytical objects, which include descriptive and analytical attributes (and families). Other concepts discussed in this section include materialized objects, granularity keys, atomic values, collections, partition keys, and bucketing/clustering keys. All these concepts are presented in the general data model (Figure 5-5). Finally, this section also discusses the concept of complementary analytical object, proper ways of joining and uniting batch and streaming analytical objects, strategies to handle dimensional data (outsourced descriptive families), and some data modeling best practices.

© FCA

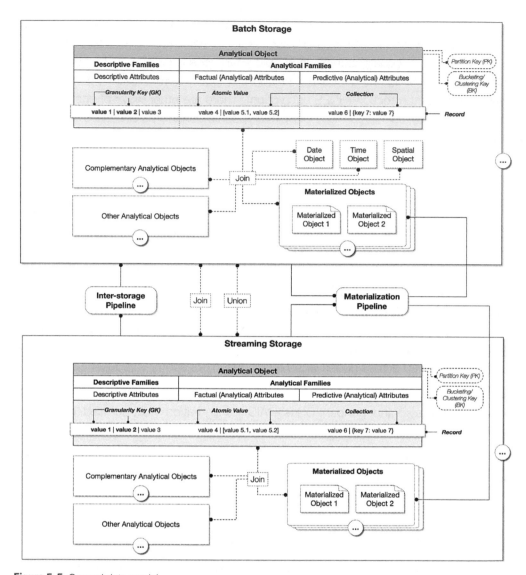

Figure 5-5. General data model.

5.4.1. Analytical Objects and their Related Concepts

In this work, an **analytical object** is defined as an isolated subject of interest for analytical purposes. Analytical objects are highly denormalized and autonomous structures that can answer queries without constantly needing to join dimension and fact tables. The benefits of full denormalized structures in terms of performance and ETL simplicity is a topic periodically discussed and evaluated by the DWing community (Jukic, Jukic, Sharma, Nestorov, & Korallus Arnold, 2017; Santos et al., 2017; Santos & Costa, 2016b; J. P.

Costa, Cecílio, Martins, & Furtado, 2011). Typical analytical objects found in organizations may include: sales; purchases; inventory management; employee vacations; employee performance; (potential) customer interactions; customer complaints; transactions during manufacturing processes. To identify an analytical object, one only needs to identify a subject of interest in a specific analytical context. Analytical objects might be found in traditional business processes or in new organizational contexts, such as social media interactions and initiatives, recommendation systems, or sensor-based decision-making, for example. An organization can identify analytical objects by either looking at the data currently being produced (data-driven), or by looking at its current goals and start collecting data to fuel these analytical objects (requirements-driven).

An analytical object includes **descriptive** and **analytical families**, as well as **descriptive** and **analytical attributes**, respectively. Families are just a logical representation for grouping related attributes, and there is no need to physically implement them in the storage system. Descriptive attributes provide a way of interpreting analytical attributes through different perspectives, using aggregation or filtering operations, for example. One can associate them with the attributes found in the traditional dimensions of a DW (Kimball & Ross, 2013). Natural keys (e.g., product code, employee code, and customer code) can also be included as descriptive attributes, if the practitioner foresees an application for these attributes (e.g., specific analyzes or update operations on records). In contrast, **analytical attributes** provide numeric values (sometimes embedded in complex/nested data structures that also contain text data) that can be analyzed through different descriptive attributes (e.g., grouped or filtered), including **factual** and **predictive attributes**. Factual attributes represent numeric evidences of something that happened in a specific record of the analytical object, and can be associated with facts in a traditional fact table (Kimball & Ross, 2013). Predictive attributes provide insights retrieved from the application of data science models and, therefore, they do not represent numeric evidences of something that happened, but rather an estimate of what happened or a prediction of what could happen in a near future. Predictive attributes are a crucial concept to adequately integrate predictive capabilities in the BDW, and can also store relevant patterns extracted from unstructured data (e.g., text, images, and video).

A **record** of an analytical object stores all the values corresponding to an event associated with that object, taking into consideration its different attributes. Descriptive and analytical attributes can contain atomic values or collections. **Atomic values** are stored as simple data types, such as an integer, float, double, string, or varchar. **Collections** store more complex structures like arrays, maps, or JSON objects. These complex and nested data structures, together with a flexible denormalized model without rigid relationships between tables, allow to explore the full potential of Big Data storage systems.

The **granularity key** is a relevant concept associated with an analytical object. The granularity key is tightly coupled with the analytical object, identifying the level of detail of the data that will be stored in each record. The granularity key of an object is defined by one or more descriptive attributes that uniquely identify a record, although this constraint does not have to be physically implemented in the storage system through a primary key, since some Big Data storage systems may not support such concept. One only needs to make sure that each record matches the granularity key of the object, which defines its level of detail.

Take as an example the analytical object "`sales`". Its granularity key can be defined solely by the unique identifier of the sales order. In this case, each record stores the general data about the sales order. The data about products sold in this order can be stored in a collection, or not stored at all, if for some reason there is no interest in carrying out that analysis. However, if the granularity key of the analytical object "`sales`" is defined by the identifier of the sales order and the identifier of the product, one record per product will be stored. There is no rigorous rule for choosing collections over redundant data stored across records, and vice versa. System orchestrators should consider their current preferences, skills, and technological or infrastructural constraints (e.g., certain querying technologies may not support collections, or the size limitations in collections may not be suitable for that context). This will depend on the implementation context. In the proposed approach, the granularity of the analytical object is never considered a limitation, nor does one apply any specific rule or guideline.

As discussed in subsection "Indexed Storage", an analytical object can be partitioned and bucketed/clustered according to specific descriptive attributes (technically, using analytical attributes is perfectly possible as well, although not as usual). The attributes that are used to partition the analytical object form the **partition key**, which fragments the analytical object into more manageable parts that can be accessed individually. This book does not provide a rigorous rule for partitioning analytical objects, but encourages system orchestrators to use time and/or geospatial attributes as the partition key (E. Costa et al., 2018), since data can be typically loaded and filtered in hourly/daily/monthly batches for specific places (e.g., cities, regions, and countries). Depending obviously on the implementation context, this is typically an adequate strategy for several contexts. Another advantage of partitions emerges in scenarios wherein data should be updated (e.g., when performing a batch update because some records were modified or were previously incorrect), which allows practitioners to recompute just the required partitions instead of the entire analytical object. In contrast, the attributes used as **bucketing/clustering key** ensure that a range of records are stored in the same group/bucket or sorted in a certain way. However, system orchestrators need to plan this strategy according to frequent access patterns requested by data consumers. The proposed approach highlights the relevance of this concept, but does not aim to provide any rule in this area, because factors may vary significantly according to the implementation context.

5.4.2. Joining, Uniting, and Materializing Analytical Objects

In the proposed approach, analytical objects can complement each other. Although there are no physical relationships implemented in the storage system, Big Data querying technologies (e.g., SQL-on-Hadoop systems) can join different datasets according to specific attributes. Therefore, an analytical object may contain in its descriptive attributes those that correspond to the granularity key (or part of it) of another object. In this case, an object is considered a **complementary analytical object** if its granularity key (or part of it) is included in another analytical object (e.g., the "customer account" object in section 6.2, whose granularity key is partially referenced by another object, and the "product" object in section 6.1, whose granularity key is fully referenced by another object). Such integration allows to associate two analytical objects through a join operation. Another type of association can be made using descriptive attributes that do not correspond to the granularity key of the analytical objects. In this case, analytical objects can be joined using regular descriptive attributes, such as a simple date, for example. A date may not define the granularity key of an analytical object, but it can be used as a join attribute between analytical objects. If many-to-many associations are identified between analytical objects, one can use collections instead, i.e., one analytical object contains a collection in its descriptive attributes that stores the association with many records of another analytical object. Once again, it must be highlighted that there is no physical relationship between analytical objects, neither is it mandatory to prepare and enrich data to create these associations between analytical objects. Technically, analytical objects can be joined by any attribute and practitioners do not need to be concerned about FK relationships and indexes. Certainly, there are many contexts in which analytical objects are analyzed independently, with no need to ever join them. However, whenever necessary, this approach offers such possibility.

At this point, a question begins to take form: *If a new denormalized approach is being proposed to solve the complexity in join-dependent data models, how can one perform efficient join operations between analytical objects if they can potentially store Gigabytes, Terabytes, or Petabytes of data?.* To answer this question, Figure 5-6 presents the process of joining analytical objects, which highlights the need to execute all the required operations in each analytical object through subqueries, or relying on efficient query optimizers to adequately and automatically process both sides of the join operation before the join itself occurs. Then, and only then, the results of these subqueries (or pre-join processing from query optimizers) are joined accordingly. This approach vastly reduces the complexity of join operations, since each subquery on both sides of the join is already as aggregated (or filtered) as possible. Figure 5-6 provides an example SQL query showing how to perform this type of join operations. If the "WITH" keyword is not compatible with the current querying and OLAP technology, one can also use subqueries in the "FROM" or "JOIN" clause. The same concepts are also valid for union operations.

This process of joining analytical objects should be applied in each join operation, including not only complementary analytical objects, but also **materialized objects** and analytical objects across different storage systems (see Figure 5-5). Since analytical objects can have a significant number of records, joining them can become a time-consuming task, even when using the join approach presented in Figure 5-6. It is in this context that materialized objects are useful and efficient. Complex and long-running queries can be materialized through the materialization pipeline (see Figure 5-5), giving origin to the materialized objects, which can be further joined with other analytical objects. The materialization pipeline also ensures materialized objects are updated with new data. Materialized objects can be stored either in the batch storage or in the streaming storage, depending on the access patterns of data consumers (e.g., using NoSQL databases for the streaming storage can provide adequate random access capabilities for specific analytical scenarios). Summarizing the concept of materialized objects, it can be concluded that they are able to store the results of time-consuming queries, increasing the performance of the BDWing system, since several data consumers can access this materialized object much faster than the original analytical objects. Consequently, materialized objects may be analogous to OLAP cubes in traditional DW environments, containing pre-aggregated data meant to be consumed in faster and more efficient ways.

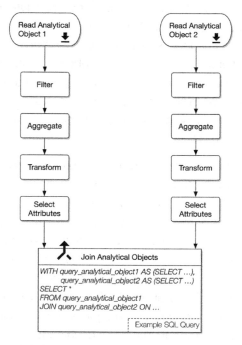

Figure 5-6. Process of joining analytical objects.

Besides join operations, this book also considers using union operations, typically useful for combining analytical objects stored in the batch storage and analytical objects stored in the streaming storage. Uniting analytical objects in different storage systems enables the visualization of batch and streaming data using a single query. Also relevant is the fact that queries can take advantage of union operators while the inter-storage pipeline does not transfer the records from a streaming analytical object to the corresponding batch analytical object.

5.4.3. Dimensional Big Data with Outsourced Descriptive Families

In certain contexts, data remains highly relational and dimensional, i.e., different analytical objects will share common descriptive families. One example is "`sales transactions`", which can be analyzed using several descriptive families such as "`customer`", "`product`", and "`supplier`", for example. These descriptive families can be included in several other analytical objects, such as "`customer complaints`", "`purchases`", "`inventory management`", among others.

As discussed previously in this section, the proposed approach allows using joins between analytical objects. However, it does not include the concept of dimensions. Typically, flat structures are preferred to avoid the cost of join operations and to achieve better performance, as demonstrated in Chapter 8. However, completely flat structures vastly increase the storage size of the BDW when compared to dimensional structures (e.g., star schema). The problem becomes worse if multiple analytical objects share the same descriptive families, because the increase in storage size can get out of control, especially if these descriptive families have a significant number of attributes. Obviously, one may be able to sacrifice storage space, which is cheaper than processing power, in exchange for better performance. Taking into consideration the insights provided in Chapter 8, it may be advantageous to use a flat analytical object that is 3 times bigger than the corresponding star schema. However, if one considers contexts with several flat analytical objects that share the same descriptive families, the BDW size can grow at a rate that the organization cannot sustain. Furthermore, there are certain contexts in which star schemas can outperform flat analytical objects (see subsection 8.2.3).

For the above reasons, and supported by the results presented in Chapter 8, this book promotes the following guidelines for modeling dimensional Big Data using the concept of **outsourced descriptive family**:

1. A descriptive family should be outsourced to a complementary analytical object if the following conditions (or the majority of them) is verified:

    a) The descriptive family is frequently included in other analytical objects (a phenomenon that is relatively similar to the conformed dimensions concept in

© FCA

Kimball's approach), avoiding extreme redundancy in the BDW, especially if the descriptive family has a considerable number of attributes. Otherwise, outsourcing frequently reused descriptive families with few attributes may not be justified;

b) The descriptive family has low cardinality, i.e., its distinct records will form a low-volume complementary analytical object that easily fits into memory, enabling the capability to perform broadcast/map joins in SQL-on-Hadoop engines (see subsections 8.2.1 and 8.2.3);

c) The frequency of data ingestion of the complementary analytical object is equivalent to the other analytical objects to which it is related. For example, if one is using the BDW to store and process streaming data from social networks, having a "user" complementary analytical object is only practical if the users' data is also streamed to the BDW as soon as a customer signs up for the social network, otherwise the BDW will suffer from problems such as the late arriving dimensions phenomenon in dimensional DWs (Kimball & Ross, 2013). If such design requirement cannot be fulfilled for some reason, then flat analytical objects represent a better option;

d) The descriptive family alone can provide considerable analytical value when analyzed independently, forming a real analytical object. For example, "customer" may serve as a complementary analytical object when outsourced from a descriptive family of another object, but it can also be used independently to measure customer performance if it contains analytical attributes related to average sales, average returns, current reviews, among other factual or predictive data.

2. Complementary analytical objects resulting from the outsourcing of descriptive families should use natural granularity keys, since maintaining surrogate keys is not practical in most BDW storage technologies, both for batch and streaming scenarios (e.g., lack of proper support for auto-increments). Searching for the surrogate keys corresponding to the natural keys flowing through CPE workloads also becomes very inefficient and unpractical, especially in streaming workloads.

3. The records of complementary analytical objects resulting from the outsourcing of descriptive families should also be designed to be immutable, whenever possible, similarly to the records of regular analytical objects. If meeting this requirement is not possible, these complementary analytical objects should be at least efficient to update or easy to recompute using a CPE workload (fully or partially using partitions), to avoid dealing with complex SCD-like scenarios. Despite this guideline, practitioners should feel free to create mutable complementary analytical objects (as well as mutable analytical objects) whenever the technologies storing the batch/streaming object support proper updates. Again, BDWing technology is evolving in this matter, and this guideline must not be seen as absolutely mandatory if performance is not severely compromised (see subsection 6.2.1 for further discussion on this topic).

4. By simply outsourcing descriptive families to complementary analytical objects, only descriptive attributes are considered. This means the resulting complementary analytical objects do not hold any analytical families and attributes, and, therefore, any analytical value. Although this is possible in the proposed approach, it somehow violates the principle that analytical objects should be autonomous structures that can answer some queries without needing any join operations. This principle will not be true for a complementary analytical object "`customer`" that will only be used to complement other analytical objects, as previously exemplified in this section. Consequently, this book encourages practitioners to use the concept of "aggregated facts as dimension attributes", described in Kimball's approach (Kimball & Ross, 2013). Although not mandatory, this technique allows practitioners to include analytical attributes (facts or predictions) in these complementary analytical objects, hence, these attributes can be used not only for filtering or labelling records, but also to perform calculations, since one is modeling an analytical object, not a traditional dimension. Using this strategy, the "`customer`" analytical object can be used to independently answer specific queries, such as *what is the average revenue generated by certain customers?*, without needing to query both the "`sales`" analytical object and the "`customer`" analytical object. Following this example, the "`customer`" analytical object can include predictive attributes, such as a cluster label based on the customer's value to the organization (see subsection 7.5.1). Obviously, in line with Kimball and Ross (2013), these pre-aggregations make the processes that make data flow to the system burdensome, but also provide more analytical value and, sometimes, eliminate the need for complex and costly queries. Such trade-offs still hold true in the proposed approach.

5.4.4. Data Modeling Best Practices

This subsection introduces some best practices that can be applied to a BDW data model, to clarify some questions that may arise during its design and implementation, including the use of null values, the preparation of spatial and temporal attributes, and the modeling of records as immutable events.

5.4.4.1. Using Null Values

Using null values in the BDW is not forbidden, and in certain occasions it is even advisable. However, there are some relevant practices that must be taken into consideration. Regarding analytical attributes, one advises the use of `null` to indicate the absence of a value, since null values are often ignored in querying, OLAP, and visualization technologies, which do not take them into account when performing aggregations on data. If numbers

© FCA

like 0 or -999, for example, are used to indicate the absence of a value, every time an aggregation is performed, filters need to be applied first to ignore these values, since they affect an average/sum calculation.

In contrast, regarding descriptive attributes with a text data type, the use of "Unknown" or "Not Applicable" is more user-friendly and appropriate when using these attributes to aggregate analytical attributes. However, there are certain data types for which using null values is still preferable (or the only solution) to indicate the absence of values in descriptive attributes; namely types such as boolean, arrays, or maps.

5.4.4.2. Date, Time, and Spatial Objects vs. Separate Temporal and Spatial Attributes

The **date and time objects** presented in Figure 5-5 include several temporal attributes that complement the analytical objects stored in the BDW. Including these attributes (e.g., "is holiday", "is weekend", "month", and "year") in the analytical objects can dramatically increase their storage size and consequently affect the stability and performance of the BDWing system. These objects are considerably small and will not significantly affect the performance of the BDW by requiring a join operation, as seen in Chapter 8.

This book encourages using date and time objects to store a vast set of temporal attributes that can be used by the analytical objects. An adequate practice would be using standard dates (e.g., "yyyy-mm-dd") and standard time representations (e.g., "hh:mm") in all analytical objects, which would then allow to join them with the date and time objects.

Following this approach, practitioners can also use several UDFs to interact with the single date or time attributes stored in the analytical objects, to create new attributes not present in the date and time objects. Extracting attributes at runtime may not impact the query execution time significantly, sometimes just showing negligible increases. Nevertheless, this book does not discourage using separate temporal attributes (e.g., "day", "month", "year", "hour", and "minutes"), quite the contrary, since they are still useful in certain contexts. One particular example is the specification of partition keys, given that, frequently, only simple data types like strings or integers can be used in the partition key. Therefore, if one needs to use "month" as the partition key, it may be necessary to have a separate temporal attribute "month". Concluding, using date and time objects or separate temporal attributes depends on the context of implementation, and system orchestrators should evaluate the most adequate solution for each context.

Regarding **spatial objects**, they prove to be significantly useful for standardizing spatial attributes across the analytical objects of the BDW, such as ensuring that a city and a country have the same exact meaning (and characteristics) throughout the entire data model. However, practitioners should be careful with large and detailed spatial objects (e.g., "`building number`", "`street name`", and "`coordinates`"), because join operations can certainly create performance bottlenecks in Big Data contexts. Therefore, one should maintain these highly detailed characteristics (e.g., "`building number`" and "`coordinates`") in the analytical object in a denormalized form, while creating less granular spatial objects like "`city`", for example, which can also include the corresponding countries in a denormalized form (see subsection 6.3.2). However, highly detailed spatial objects are acceptable in scenarios with a predictable growth, because the number of records they can have is known or expected *a priori.*

5.4.4.3. Immutable vs. Mutable Records

As previously discussed, this book encourages practitioners to model analytical objects as a set of immutable events. As Marz and Warren (2015) discuss, simpler implementations can be achieved by eliminating the complexity associated with update operations, which can sometimes raise concurrency issues. This modeling style will probably suit most analytical scenarios in organizations, since the granularity of each analytical object can be rethought to treat each record as an immutable event.

Take as an example an analytical object to store customer complaints (Figure 5-7). A certain organization knows that a customer complaint has several states over time. A possible approach, which allows the records to be updated, is to have one analytical object that stores a customer complaint in each record. When a recently opened customer complaint arrives at the BDW, it is stored in a record with the status "`open`", not having a due date yet. Meanwhile, this record will have to be updated when the customer's complaint is "`finished`". Another approach is to model the analytical objects according to a set of events related to customer complaints. When a recently opened customer complaint arrives at the BDW, a record is created containing the status and the date associated with that status. When the status of the customer complaint changes, new data arrives at the BDW, and a new record for each state change is stored. This second approach ensures that each record is immutable, eliminating the need for update operations.

Despite the fact that queries need to be structured in different ways, the two analytical objects presented in Figure 5-7 can answer the same analytical questions. Furthermore, one can argue that the immutable analytical object is more oriented towards *ad hoc* querying, wherein data consumers can discover relevant patterns and delays among processes related to customer complaints. However, the proposed approach does not forbid the use of mutable analytical objects, considering that practitioners plan the BDW's technological infrastructure taking into account the random access trade-

© FCA

offs and limitations of the several technologies presented in section 5.3. Modeling analytical objects as a set of immutable events is a suggestion, not a rigorous rule, since updates can be performed on storage systems that adequately support random access operations, as previously discussed in subsection "Indexed Storage" and further explored in subsections 6.1.3, 6.2.1 and 6.2.4. As previously discussed, technology is constantly evolving, and these trade-offs or limitations may not be an issue in certain implementation contexts. The proposed approach does not aim to restrict any use of specific functionalities, giving practitioners an adequate flexibility regarding data modeling. However, this book highlights the need to ensure that the logical components, data flows, infrastructure, and data model are properly integrated and coordinated to serve the business goals.

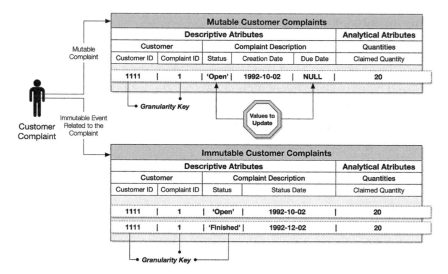

Figure 5-7. Example of immutable and mutable records.

5.4.5. Data Modeling Advantages and Disadvantages

This modeling approach based on denormalized and nested data is a crucial step to achieve a flexible storage in the BDW. When compared to relational data modeling approaches found in traditional DWs, this work exchanges less redundancy and smaller DW sizes for the following advantages:

1. Better performance during query execution, due to the lack of constant join operations between dimensions and fact tables imposed by traditional dimensional and 3NF data models.

2. A flexible denormalized model without the need to perform complex surrogate key maintenance and lookups for each insert, allowing simpler and more efficient batch and streaming CPE processes, by avoiding known-problems such as SCDs and late arriving dimensions (especially in streaming scenarios).

3. A focus on modeling analytical objects as a set of immutable events and, therefore avoiding dealing frequently with concepts such as SCDs (Kimball & Ross, 2013). However, as explored in subsection 6.2.1, this does not mean that mutable objects are forbidden, and when using them, some of the SCDs considerations still hold true.

4. Avoids other traditional dimensional data modeling, ETL, and DW maintenance problems like having to consider several types of dimensions (e.g., mini dimensions, junk dimensions, shrunken dimensions, and bridge tables), which in Big Data contexts are arguably unnecessary, since saving storage space and achieving less-redundant data models may come at the cost of spending a considerable amount of time in data modeling, implementing ETL processes, and maintaining the DW (not to mention performance costs), which may be a compelling reason why, nowadays, practitioners pursue more flexible analytical contexts. Consequently, despite some data redundancy, in several contexts, the proposed approach provides simpler data models than a dimensional or 3NF DW, reducing the time needed from collection to analytics.

5. Highlights nested structures as relevant constructs in certain BDW data models and applications, which can be useful in certain contexts (see Chapter 6 and Chapter 9), such as when storing geospatial objects for intensive geospatial analysis, and solving many-to-many relationship issues typically found in relational databases (e.g., a customer complaint may have several responsible employees, which are also responsible for several customer complaints).

Nevertheless, the proposed data modeling method has some characteristics that may be considered disadvantageous when compared to the aforementioned methods for designing DWs, which include:

1. The total size of certain BDWs (typically the ones whose data sources are highly dimensional with frequently reused dimensions) may increase drastically due to extreme denormalization, hence why the approach introduces the concept of date/time objects, spatial objects, complementary analytical objects, and outsourced descriptive families. Consequently, practitioners should take into consideration the guidelines provided in subsections 5.4.4.2 and 5.4.3, as well as the data models explored in Chapter 6, mainly in sections 6.1 and 6.3, as the original data sources tend to be highly dimensional, with the same dimensions being reused frequently by different business processes/analytical subjects. Without these strategies, the resulting BDWs would be significantly larger than the DWs based on star schemas or 3NF data models. Nowadays, storage size is cheap, but may often lead to unnecessary concerns and costs regarding systems administration, which can be prevented by using the constructs discussed above, whenever practical and applicable.

2. If the data source fueling an analytical object is based on a relational database, the CPE workloads for that object may need to include a considerable amount of join operations, either being performed in the source (as a SQL query for example), or in the technology supporting the workloads. However, in Big Data contexts, many of the data sources are non-relational (e.g., sensor data, NoSQL databases, spreadsheets, XML files, and JSON files), making the proposed method for data modeling significantly more compelling and simpler for BDWs.

6 BIG DATA WAREHOUSES MODELING: FROM THEORY TO PRACTICE

INTRODUCTION

After presenting the general data modeling method in section 5.4, this chapter explores its use in several BDWing contexts, since more practical examples and real-world applications may be required for practitioners to master some of the proposed data modeling guidelines. Consequently, this chapter aims to provide several examples of BDWing applications using the proposed data modeling method, to clarify some of the guidelines provided previously, and to evaluate their suitability in a broader scope of analytical applications focused on: traditional enterprise setups with human resource management, purchases, sales, promotions, goods returns, inventory management, and production process; financial market; retail; code version control systems; media events (broadcast, printed and Web news); and air quality measurement systems.

6.1. MULTINATIONAL BICYCLE WHOLESALE AND MANUFACTURING

As already seen, Big Data can be defined as data whose characteristics impose severe difficulties for traditional DWing platforms. Frequently, there may be a misconception regarding the need to satisfy all Big Data characteristics when deploying a BDW, such as the need to process vast amounts of unstructured data arriving at theoretically unlimited velocities. However, in this section, we will discuss how a BDW can be modeled to encompass traditional business processes like human resources management, sales, purchases, production, among others.

Traditional DWs have long been the backbone of analytics over traditional and structured business processes; this section provides a way of modeling such complex scenario in a BDW created using the proposed approach, to provide more data modeling simplicity, less ETL effort without complex dimension maintenance and SK lookups, and more processing efficiency by reducing the constant need to join several tables. Such benefits can attract organizations that are starting their analytical platforms based on shared-nothing and open source Big Data technologies, as well as organizations looking to replace their expensive DW appliances or limited relational databases.

For this example, we use the Adventure Works database, a relational OLTP database from a fictitious company that manufactures and sells bicycles, included as part of the Microsoft SQL Server samples (Microsoft, 2018b). This database has a relatively complex schema that covers a wide spectrum of business processes and entities (e.g., employees, vendors, customers, stores, departments, products, work/production orders, purchases, sales, and inventories). The complete representation of the Adventure Works database is available in (Dataedo, 2017).

After applying the data modeling method, the resulting BDW data model can be seen in Figure 6-1. It contains seven analytical objects ("`employee history`", "`sales line`", "`product review`", "`product vendor history`", "`purchase line`", "`product inventory`", and "`work order`"), three complementary analytical objects ("`product`", "`vendor`", and "`special offer`"), one date object, one time object, and two spatial objects ("`city`" and "`territory`"). Descriptive attributes are divided into descriptive families, whereas analytical attributes are divided into analytical families, when applicable. Analytical objects can also contain outsourced descriptive families that are linked to a complementary analytical object through a unique identifier (granularity key) of that object, identifying a specific record. Several of these constructs and design guidelines, which were already discussed in section 5.4, are detailed and exemplified here, not only for this specific example, but also for the other BDW examples in the following sections.

The data model presented in Figure 6-1 sometimes omits certain attributes of the original Adventure Works database, in order to simplify its presentation in this work;

these omissions include the attributes in the "`header`" analytical family of the "`sales line`" analytical object due to its similarity with the "`purchase line`" analytical object or the attributes from the "`customer`" and "`sales person`" descriptive families of the "`sales line`" analytical object, due to the wide spectrum of available attributes (different practitioners may choose to incorporate different attributes). Therefore, the main idea is to exemplify the modeling approach without extensively enumerating the attributes.

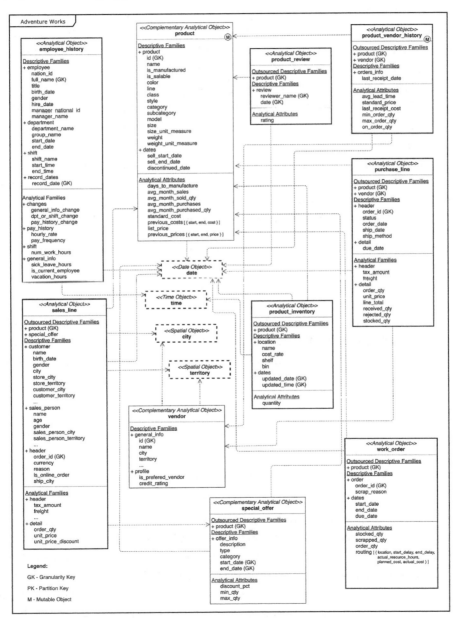

Figure 6-1. Adventure Works BDW data model.

6.1.1. Fully Flat or Fully Dimensional Data Models

The example in Figure 6-1 demonstrates the use of outsourced descriptive families and complementary analytical objects (subsection 5.4.3), using them to overcome extreme redundancy and storage size increase. By revisiting the arguments for using these concepts in subsection 5.4.3, we highlight the following:

1. "`product`" is an adequate candidate for a complementary analytical object because its attributes would otherwise appear repeated in several analytical objects, since in this context a product is a core business entity. A "`product`" object allows to standardize the products' information across the BDW, and since in this context new products are not added rapidly, this is an adequate design choice, because it will not affect join performance, as broadcast/map joins will remain efficient for a long time. The "`product`" object by itself holds a significant analytical value, distinguishing it from a traditional dimension to avoid redundancy, as one can be interested in analyzing several metrics regarding products, without needing any additional analytical objects. Therefore, this is a valuable construct in the approach, and it resembles the concept of "aggregated facts for dimensions" from Kimball and Ross (2013). In subsection 6.1.3, we will detail how this concept can be implemented.

2. For the same reasons, "`vendor`" is also an adequate complementary analytical object that serves two outsourced descriptive families from the "`product vendor history`" and the "`purchase line`" analytical objects. However, in contrast to "`product`", "`vendor`" does not have any evident analytical attributes, although "`is preferred vendor`" and "`credit rating`" could be considered analytical attributes as well. The proposed approach offers this flexibility due to the denormalization process, allowing the execution of aggregate functions over any attribute present in the analytical object without involving any kind of join operation. Moreover, as explained above, other analytical attributes can be created (e.g., average monthly purchases). Another relevant consideration is the fact that "`vendor`" is related to the spatial objects, so we can conclude that, since there is no need to define FKs in BDWs created using the proposed approach, objects in the data model can be flexibly joined, as long as there are common unique identifiers between them (simple or composed).

3. "`special offer`" is considered a complementary analytical object, although it is only related to the "`sales line`" analytical object and, therefore, it does not necessarily serve the purpose of avoiding extreme redundancy. However, theoretically, it represents a standard analytical object that happens to be joinable with the sales information by a unique identifier. Consequently, as seen in subsection 5.4.2, two analytical objects can be joined together, being the designation of complementary analytical object assigned to the object whose granularity key (or part of it) is included in other objects, in this case making "`special offer`" a complementary analytical object of "`sales line`".

4. Other potential candidates for complementary analytical objects could be the "`employee`" and "`customer`" objects. Regarding a possible "`employee`" complementary analytical object, there is employee information in the "`employee history`" and "`sales line`" objects but, in this model, one can consider that only a subset of the employee attributes are relevant for each analytical object, thus denormalization and redundancy is appropriate and, therefore, there is no need for a complementary analytical object integrating the employee information. In the case of the "`customer`" analytical object, since customer information only appears in the "`sales line`" analytical object, there is no apparent need for a complementary analytical object that can be shared by other analytical objects, being the level of denormalization presented in Figure 6-1 appropriate for this context. However, creating a "`customer`" analytical object is possible and sometimes encouraged, as can be seen in the data model depicted in section 6.3.

6.1.2. Nested Attributes

Nested attributes are a valuable construct in the proposed modeling method, as they offer considerable flexibility and a new set of analytical possibilities. As can be seen in Figure 6-1, considering the "`work order`" analytical object, one can observe that although this object stores information at the work order level, the routing attribute stores more granular information at the work order route level, detailing the several production steps involved in a specific order. This allows a broader range of ad hoc queries to inspect routing information, without needing heavy drill across operations.

As discussed in subsection 8.2.4, lambda or explode functions can be used to explore nested data. Nested attributes are also used in the "`product`" complementary analytical object to store the history of prices and product costs. These attributes are arrays of structs/rows (or similar data structures), and can serve to analyze price/cost history of a specific product, again, without the need to join tables. These constructs are powerful for ad hoc exploration of data, but require some attention when performing heavy aggregations or filtering operations based on nested values, as seen in subsection 8.2.4. Another relevant aspect to consider is the size of the collections, as they are not meant to grow rapidly, due to the fact that some Big Data technologies may present limitations when performing insert, read, or update operations on large nested attributes. Consequently, they are more adequate in scenarios wherein practitioners can estimate their initial size and potential growth.

6.1.3. Streaming and Random Access on Mutable Analytical Objects

As noted in Chapter 5, this book encourages the storage of immutable events, not only because some of the core concepts of the approach take inspiration from the Lambda Architecture, but also due to some current limitations of Big Data storage technologies when performing update operations (e.g., HDFS/Hive). However, this guideline does not prevent practitioners from modeling and implementing mutable (complementary) analytical objects. In this subsection, we will discuss how mutable objects can be incorporated in a BDW, considering "`product`" and "`product vendor history`" as examples.

As previously noted, some of the analytical attributes from the "`product`" complementary analytical object resemble the concept of aggregated facts for dimensions (Kimball & Ross, 2013) (e.g., "`avg month sales`" and "`avg month sold qty`" attributes). However, without proper support for update operations, each month this analytical object would need to be completely reconstructed to store the new monthly values. In contrast, if needed, as discussed in section 5.3, practitioners may opt for storage systems that are suitable for random reads and writes. When choosing a NoSQL database, for example, one does not need to recompute the "`product`" object, just update the average monthly metrics for each product.

The proposed approach assumes that this type of design choice follows the streaming data flow shown in Figure 5-2, because this book only suggests NoSQL databases for the streaming storage component, not the batch storage component. However, in this case, the updates happen in relatively large batch intervals, which may or may not be supported by streaming technologies depending on the CPE workload execution frequency (e.g., every time a customer purchases something, each day, or each month). Such assumption forces these analytical objects to be stored in the streaming storage component, regardless of whether the CPE workload is based on a batch or stream processing. This is a design choice in the proposed approach, as the batch data flows remain considerably similar to constantly inserting/updating values on a streaming analytical object stored in a NoSQL database.

Nevertheless, with the rapidly evolving Big Data technological landscape, support for update operations and ACID transactions is an issue for several storage technologies, and Hive is no exception. Therefore, if practitioners choose a Hive transactional table to store products data, this scenario can be adequately supported by the batch storage component, without needing to store the "`product`" analytical object in a NoSQL database (streaming storage). Transactional tables have been significantly optimized in Hive version 3 (Apache Hive, 2018), hence they are a relevant feature to explore in future prototypes and production systems. Consequently, nowadays, practitioners do not have to choose NoSQL databases to adequately perform random insert/update operations with moderate frequency.

The context for the "`product vendor history`" is almost identical to the previous one. In contrast, "`employee history`" is an example of how a potentially mutable object can be transformed into an immutable object, since each time some employee data changes (e.g., personal information, department, shift, or salary), a new record is created, which allows to analyze employee history in significantly flexible ways.

6.2. BROKERAGE FIRM

The financial sector has increasingly been considering the adoption of Big Data techniques and technologies as part of the Fintech phenomena (Gai, Qiu, & Sun, 2018). A brokerage firm that facilitates financial securities trading can represent an appealing application context for a BDW, given that it stores and processes vast amounts of daily market and news data, as well as trading and watching data from several securities related to multiple brokers and customer accounts. Consequently, in this section, we model a BDW for a fictional brokerage firm depicted in the TPC Benchmark E (TPC-E) (TPC, 2018), which thoroughly details a concurrent transactional database system adapted to financial brokerage contexts.

For this book, we transformed the TPC-E data model into a BDW data model using the proposed approach (Figure 6-2). The brokerage firm BDW data model is presented in a simplified manner, to avoid the repeated constructs already discussed in this chapter, therefore, some (complementary) analytical objects are not detailed at the family or attribute level.

6.2.1. Unnecessary Complementary Analytical Objects and Update Problems

In the BDW data model depicted in Figure 6-2, there are three complementary analytical objects: "`customer account`", "`broker`", and "`security`". In this example, "`customer`" and "`company`" could theoretically be included as complementary analytical objects, but due to their lack of isolated analytical value for this specific context, and to the frequency in which they appear related to other objects, they were not considered as complementary analytical objects. Hence, we opted for some denormalization steps: "`customer`" data appears denormalized in the "`customer account`" object; "`company`" data appears denormalized in the "`news`" and "`security`" objects.

The above design decision also means that the "`watch list`" analytical object, which in the original TPC-E model is related to the "`customer`" table and not to the "`customer account`" table, needs to be indirectly joined with the "`customer account`". In this case, for example, to retrieve customer information associated with specific watch list data one needs to perform a left outer join retrieving the customer information from its last customer account. Moreover, if there is a change in some attribute related to the customer and not the customer account, one needs to choose an update strategy:

1. Replace the values across all the related customer accounts by scanning the entire analytical object or several partitions.

2. Insert a new record for each customer account with the updated values.

3. Update only the last customer account.

4. Update only the customer accounts when a new account is inserted, given that the customer created the accounts before this update and such information might be somehow valuable for business analysis (immutable events strategy).

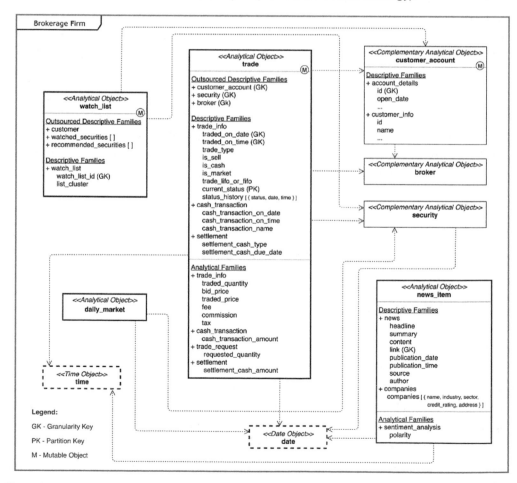

Figure 6-2. Brokerage firm BDW data model.

If practitioners find this design approach suitable for their use cases, the same can be implemented to offer more simplicity in CPE workloads, otherwise a new complementary analytical object "`customer`" can also be created, as the approach provides this flexibility by delegating some design decisions to practitioners according to their implementation's specificities. Regarding update operations on complementary analytical objects, design choices are often influenced by the adoption of a specific technology (see section 5.3 and subsection 5.4.4.3), due to their random access or batch update capabilities. However, some of these choices and challenges are also somehow related to the concept of SCDs (Kimball & Ross, 2013), as some of the underlying challenges of updating denormalized dimensions resemble the challenges of updating complementary analytical objects, due to data redundancy (scanning vast amounts of data to update certain values) and history maintenance.

6.2.1.1. The Traditional Way of Handling SCD-like Scenarios

Several strategies from multiple SCD types (e.g., SCD types 1, 2, and 3) can also be applied to complementary analytical objects, but one needs to consider that the proposed approach does not have the concept of SK and, therefore, practitioners should rely on the originally defined granularity key, as well as modification dates and flags to indicate the current/active records, when needed, in order to appropriately join analytical objects, which creates a slightly more complex granularity key (granularity key information on subsection 5.4.1). In this example, "`customer account`" would not have a simple "`customer account id`" as granularity key, but a complex granularity key such as "`customer account id`", "`insert date`", "`expiration date`", and "`is current`", for example.

6.2.1.2. A New Way of Handling SCD-like Scenarios

Since data engineering practices are constantly evolving, practitioners of the Big Data community often share insightful techniques to handle complex scenarios. In this context, Beauchemin (2018) shared a technical article regarding a useful perspective on functional data engineering. In this article, two techniques related to handling SCDs are highlighted, which can be applied within the context of the approach proposed in this book, namely in cases where update problems on complementary analytical objects arise:

1. **Dimension snapshots** (in the context of this book, complementary analytical objects snapshots) – a new partition is created at each ETL execution (equivalent to a data CPE process in this book), containing a snapshot of the dimension (in the context of this book, a snapshot of the complementary analytical object). Consequently, the complementary analytical object becomes a collection of snapshots that allows inferring its state at a given moment in the past. In this way, one can join analytical

objects with any snapshot of any complementary analytical object that follows this technique. Since complementary analytical objects are typically smaller than analytical objects, and since storage is cheap nowadays, this technique can be feasible in many contexts. However, for large complementary analytical objects, avoiding this technique is probably advisable.

2. **Embed the attributes change history in nested structures** – Beauchemin (2018) states that another modern technique is to store the change history in nested data types. In the context of this book, this will enable practitioners to track changes to attributes without changing the granularity of the complementary analytical object. In this way, one does not need to create several attributes to track change history, since a complex and nested structure can be used.

6.2.2. Joining Complementary Analytical Objects

As has been discussed throughout this book, the approach considers all tables analytical objects, which can be complementary, or not, depending on whether or not they contain descriptive attributes outsourced from other objects. Frequently, as seen in this brokerage firm, complementary analytical objects may resemble traditional dimensions, despite the fact that one encourages practitioners to provide analytical attributes for these complementary objects. This is the case for the "`broker`" and "`security`" objects in this example. Considering the guidelines provided in subsection 5.4.3, for BDW data models with significantly large complementary analytical objects created with the purpose of supporting outsourced descriptive families, if interactive query execution is a priority, one should consider denormalizing data even further. This task can be accomplished by including attributes from the "`security`" object in the "`trade`" object for example, taking into consideration the data model of this brokerage firm. This may be the case for the "`security`" complementary analytical object, which can grow significantly large depending on the securities being traded in this context.

6.2.3. Data Science Models and Insights as a Core Value

One of the main design concerns in the proposed approach involves closing the gap between data science models/results and the BDW data structures that store data for later use. Throughout this book, we discussed this topic on several ocasions (see subsection 5.2.2 and section 7.5). For the brokerage firm, we can apply the concept of predictive attributes to make data science results available to other analytical applications (e.g., dashboards, *ad hoc* querying, custom-made applications, and simulations). Such examples may include: i) the "`recommended securities`" and the "`list cluster`" in the "`watch list`" object, which can be derived from a recommendation engine and a clustering algorithm respectively; and ii) the "`polarity`" attribute from the "`news item`" object, which may be the result of a sentiment analysis process that classifies a news item as being positive or negative, to enrich the decision-making processes that the BDW can support.

In contexts where custom-made applications may need to access data stored in the BDW, such as a brokerage firm website that recommends securities to millions of customers based on the "`watch list`" recommendations, the "`watch list`" analytical object becomes an adequate candidate for a streaming analytical object that is stored in a NoSQL database. That decision offers adequate random access to millions of concurrent users, a use case wherein NoSQL databases thrive (strategy already discussed in subsection "Indexed Storage", section 5.3, and subsection 6.1.3).

6.2.4. Partition Keys for Streaming and Batch Analytical Objects

Considering this financial brokerage context, the "`trade`" object is noticeably the analytical object where most of the decision-making process will be focused. Analyzing a stream of trading data can provide significant business value, accelerating the decision-making process in various ways. However, a trade follows different stages (e.g., request, cash transaction, and settlement), and as modeled in Figure 6-2, it may have different attributes filled in depending on its type (e.g., cash or margin trade).

One of the constructs that can be used in this context is the partition key. By using this construct, practitioners can use the same analytical object to store both batch and streaming records, in this case, trading data. For example, if one partitions the "`trade`" object using the "`status`" attribute (or any other attribute available in the transactional system indicating different states of the trade), both batch and streaming data can be stored in the same analytical object and in the same storage technology (e.g., Hive), wherein the trade can be constantly updated until it reaches a state of completion. By using different Hive partitions to divide batch and streaming records of the same table, one can have different schemas for each partition, which means that some attributes of the "`trade`" analytical object may only be included in specific partitions, depending on the state of the trade (e.g., requested or settled). This is only possible for technologies that can handle schemas defined at the partition level.

This capability also means that frequent use of update operations (e.g., Hive transactions) can be restricted to streaming partitions, because once the trade reaches completion, the chances of it being updated are slim. This demonstrates the flexibility of the proposed approach, which allows a seamless integration between batch and streaming data, and efficient ways of conducting update operations, even though it encourages the modeling of immutable objects whenever possible. However, in this

© FCA

case, to provide a timely and interactive analysis, the "`trade`" analytical object can be made mutable without sacrificing efficiency, due to technological evolutions like Hive's transactions (Apache Hive, 2018).

6.3. RETAIL

In this section, we provide an example of a BDW that supports a retail organization derived from the TPC Benchmark DS (TPC-DS) (TPC, 2017a), with store, catalog, and Web sales. This section provides some specific details regarding retail contexts that may be useful for practitioners, and that were possibly overlooked in the Adventure Works BDW (see section 6.1), since it represents a broader organizational context. The retail BDW data model presented in Figure 6-3 comprises various analytical objects (including complementary) in a highly dimensional model, focusing on sales, returns, promotions, customers, items, and warehouses.

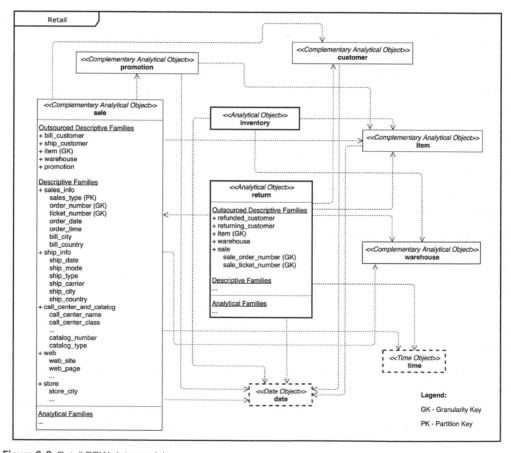

Figure 6-3. Retail BDW data model.

6.3.1. Simpler Data Models: Dynamic Partitioning Schemas

Similarly to the concepts demonstrated in subsection 6.2.4, the retail BDW data model presented in Figure 6-3 also uses the partition key to provide simplicity and agility when collecting, preparing, and enriching the data that flows to the BDW. However, considering this example, one does not use the partition key and dynamic partitioning schemas to simplify batch and streaming analytics in the same analytical object, but rather to provide simpler data models. By using different schemas for different partitions, one can efficiently store what would otherwise be three separate analytical objects into just one, i.e., store, catalog, and Web sales into the "`sale`" analytical object partitioned by "`sales type`". Each partition can have different attributes, which provides a centralized and efficient way of storing each type of sales. This phenomenon also happens for the "`return`" object, as it is almost identical in structure when compared to the "`sale`" object, according to this specific retail context. This is only possible for combinations of storage and processing technologies and file formats that can handle different schemas defined at the partition level. Otherwise, practitioners will need to consider the following: i) use the same schema for the entire analytical object and use null values (or unknown values) when specific partitions do not have values for specific attributes; or ii) do not use different schemas for different partitions and, instead, use various analytical objects as needed (three objects in this case: store, catalog, and Web sales).

Furthermore, in this example, "`sale`" is considered as a complementary analytical object, since the "`return`" object includes the granularity key of the "`sale`" object in its descriptive families, due to the fact that a return is related to a "`sale order/ ticket number`" and an "`item`". Such relationship may resemble scenarios in which practitioners use degenerate dimensions for drilling across fact tables, first aggregating the two result sets, as much as possible, and then combining the results, as discussed in subsection 5.4.2.

6.3.2. Considerations for Spatial Objects

According to the proposed approach, *a priori* designed spatial objects are not mandatory. However, as seen in the previous data models, they are encouraged in predictable scenarios. Considering this retail context, despite the fact that customers have specific addresses, sales are frequently not billed nor shipped to the default customer address and, therefore, they end up being also attached to the sale itself, not only to the customer. It is possible, and perfectly plausible to include a spatial object (e.g., city) in the data model depicted in Figure 6-3, but, for this example, we show that it is not mandatory to have one. One may choose to perform the analysis at the city and country level only, i.e., without other standardized spatial attributes across the BDW (e.g., county, region, and

© FCA

continent), which makes the effort of having to join the "`sale`" or "`customer`" analytical objects with a "`city`" spatial object with more attributes almost useless.

Choosing adequate attributes suitable for analyzes should always be a relevant consideration (see Figure 6-3), and it will influence the use of wide spatial objects with several attributes or a few denormalized attributes in the analytical objects. Both possibilities are adequate for this context, but this example only serves to highlight that in specific contexts spatial objects may not be particularly useful. Furthermore, one aspect that practitioners should take into consideration is avoiding large spatial objects (e.g., denormalized hierarchies ranging from building numbers to country names). In this case, some of the more granular geospatial information can be contained within a descriptive family of the analytical object (e.g., building number and building type), and the less granular information can be stored in the spatial object (e.g., city and country).

6.3.3. Analyzing Non-Existing Events

Considering a traditional DW, if one uses a "`customer`" transactional table to directly load a "`customer`" dimension, the DW will be able to answer queries such as the following: "`which customers have not returned a single item?`". However, considering a BDW with a fully denormalized analytical object "`return`", such analysis would not be possible; that is why practitioners have the option of using complementary analytical objects like "`customer`". The same consideration holds true for spatial objects, as one may want to analyze the cities in which the organization did not sell any item. Consequently, for such analytical use cases, practitioners should definitely consider complementary analytical objects, as well as date, time, and spatial objects, since fully denormalized analytical objects only store the events (records) that actually occur.

6.3.4. Wide Descriptive Families

Previously, in subsection 6.3.2, we highlighted the importance of adequately choosing the attributes that are relevant for the expected analyzes. This does not imply that queries that will be submitted to the system have to be known in advance. Nevertheless, frequently, there are certain attributes that are considered irrelevant for the analytical use cases of the BDW being implemented. In such cases, adequately choosing the attributes allows smaller descriptive families, which is relevant when using fully denormalized structures. With larger descriptive families, more redundant data would be stored throughout several records, instead of just one or few attributes that allow for join operations with complementary analytical objects.

Regarding the retail context illustrated in this section, the "store" descriptive family from the "sale" object can, in theory, hold a considerable number of attributes. However, certain attributes may be considered irrelevant depending on the analytical use cases, such as the store's "GMT offset" or "tax percentage", if the decision-making process of the organization does not consider such information. Consequently, narrow descriptive families should be used whenever possible without sacrificing analytical value. Despite this guideline, if wide descriptive families are mandatory for a specific case, columnar file formats (e.g., ORC and Parquet) with compression techniques can be an efficient way of storing analytical objects with hundreds or thousands of columns.

Furthermore, if needed, one can create a "store performance" complementary analytical object related to sales, outsourcing the "store" descriptive family, since this object would provide significant analytical value at the store level, including several ratios between number of workers, floor space, and sales numbers. The flexibility of the approach regarding dimensional data allows to delegate some design decisions to practitioners, depending on the intended analysis and data characteristics.

6.3.5. The Need for Joins in Data CPE Workloads

Considering the TPC-DS data model (TPC, 2017a), information like customer demographics, customer household demographics, customer income, and customer address appears related using FK relationships between the dimensions containing this information and the "customer" dimension. In the BDW presented in Figure 6-3, all this information is denormalized into the "customer" complementary analytical object. Again, if one needs to answer queries such as *is there any customer demographic class in which the organization does not have any customer?*, this design choice is not appropriate, and the "customer demographics" descriptive family inside the "customer" object will need to be outsourced to a complementary analytical object. However, we assume this is not the case in the retail context.

Considering this denormalization process, with a "customer" complementary analytical object that includes demographics, household, address, and income information, at first glance, the data CPE process may appear simpler than maintaining several separate dimensions. However, the degree of simplicity depends on the transactional source that fuels the "customer" object:

- If the transactional source is a relational database whose information comes from several tables, then the data CPE workload corresponding to the loading and refreshment of the "customer" object will need to perform several joins to achieve a fully denormalized structure;

- In contrast, regarding large-scale retail scenarios using NoSQL databases to support the vast amount of transactions being generated, this data may arrive at the BDW already denormalized (e.g., column-oriented and document-oriented NoSQL databases), representing the opposite situation and providing a high degree of simplicity without the need to perform join operations, which considerably simplifies the data CPE workload.

6.4. CODE VERSION CONTROL SYSTEM

The software industry is under constant evolution, and open source or subscription-based remote version control systems such as GitHub have become a pillar of current software management and dissemination. GitHub is one of the main collaboration platforms for software development, whose activity generates vast amounts of data. In this section, we explore the GitHub public dataset available on Google BigQuery (Google, 2018) comprising 2.9 million public software repositories, to model a BDW that supports decision-making process concerning the activity and metrics of these repositories' commits and content in large-scale environments. The BDW data model illustrated in Figure 6-4 includes the "commit" and "repository" analytical objects, where the latter complements the former, and also includes the date and time objects.

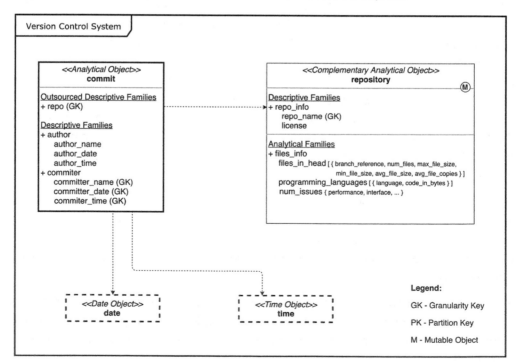

Figure 6-4. BDW data model for a code version control system.

The "`commit`" analytical object stores data about the commits that have been made to the several repositories, including information concerning the author and the committer. This analytical object does not contain any relevant analytical attribute and, therefore, count operations will be the primary focus of analysis. The "`repository`" complementary analytical object stores information regarding the current state of the 2.9 million public repositories, including the license, an array containing the information of several files for each branch, an array containing the code (in bytes) of each programming language in the repository, and the number of issues classified by type (possibly extracted by scrapping and mining the text from the issues page of each repository, for example).

Both the "`commit`" and the "`repository`" objects can be implemented as streaming analytical objects, in which they are updated as soon as each commit or any other file activity takes place. However, due to the chosen data model, the streaming implementation may differ, since the "`commit`" object is an immutable append-only object, in which each commit originates a new record, whereas the "`repository`" object is a mutable object, because the nested analytical attributes should be updated (e.g., code in bytes and number of files) instead of originating a new record. Consequently, the "`repository`" object can be implemented using a NoSQL database with adequate support for fast random access to nested objects. Hive transaction tables can also be an option depending on specific implementation details such as update frequency, latency requirements, and update throughput, as previously discussed.

6.5. A GLOBAL DATABASE OF SOCIETY – THE GDELT PROJECT

The GDELT project comprises an open database that monitors worldwide broadcast, print, and Web news, identifying the people, locations, organizations, topics, sources, emotions, among many other information about news (GDELT, 2018). The data model presented in Figure 6-5 represents a BDW to support decision-making processes using worldwide event data from the GDELT project, which is composed of date and time objects, a "`city`" spatial object (also including denormalized data about the countries of the stored cities), and an "`event`" analytical object. This analytical object is responsible for storing news/events, with data regarding the event and the actors involved in it. The actors' data regarding name, city (attribute related to the spatial object "`city`"), group (e.g., United Nations or World Bank), ethnic and religion information, geocoordinates, among others, is stored in a complex data type (e.g., row or struct) for organization purposes, which can also contain other complex data types (e.g., "`religions`" and "`types`" arrays). Consequently, the "`event`" object supports several analytical applications to process and analyze worldwide news/events.

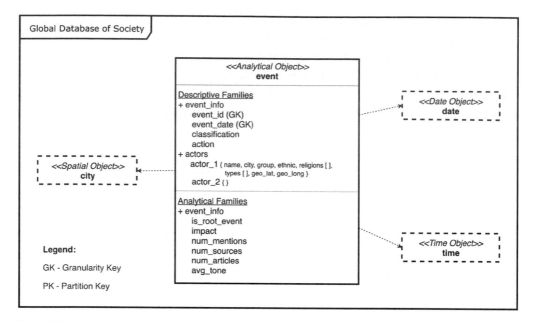

Figure 6-5. BDW data model for the GDELT project.

6.6. AIR QUALITY

The final BDW example in this chapter focuses on air quality analysis through sensors spread across different locations. The example presented in this section is based on the open air quality platform (OpenAQ, 2018). The BDW data model depicted in Figure 6-6 integrates a spatial object "city", date and time objects, and a "measurement" analytical object corresponding to the measured value of a specific parameter from a specific location, date, and time. The "measurement" analytical object has geospatial coordinates which are not present in the spatial object. This is a design choice that is always encouraged, due to the high cardinality of geospatial coordinates. Consequently, space is broken down into levels of detail, the lower levels being typically stored in spatial objects, whereas higher levels are stored in the analytical object, as already explored in subsection 6.3.2.

Real-time aggregations on sensor data are an adequate use case for specific technologies like Druid, a columnar storage that provides aggregations and indexing at ingestion time. Such design and implementation choice can fuel a "measurement" analytical object modeled at a higher level of detail, since, for example, the "value" attribute can be an average of each minute, instead of the raw sensor readings produced each second. Besides Druid, this scenario can also be supported, for example, by a Spark Streaming CPE workload using window operations or micro batch aggregations, storing the resulting data in the streaming storage system of the BDW.

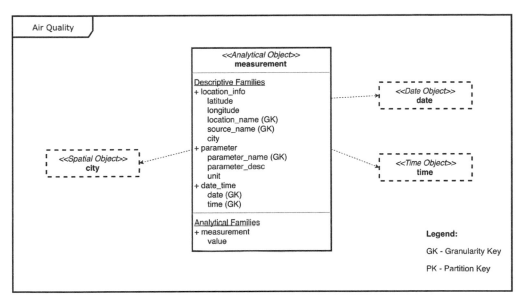

Figure 6-6. BDW data model for air quality analysis.

Nevertheless, when using Druid (or similar technologies), one should pay attention to the specificities of the data models required by these technologies, because, for example, Druid currently handles descriptive and analytical attributes in fully denormalized structures, that do not correspond entirely to the data model presented in Figure 6-6. However, as seen in this work, the proposed approach for BDWing is relatively flexible and, if that is the case, practitioners can adopt a fully denormalized "measurement" analytical object without spatial, date, and time objects. After these considerations, this section can be seen as a collection of insights that practitioners can use to design streaming analytics on sensor data, particularly for air quality analysis in this case, but also with further applications for other sensor-based analytical workloads.

7 FUELING ANALYTICAL OBJECTS IN BIG DATA WAREHOUSES

INTRODUCTION

One of the most laborious stages in the implementation of DWs, whether they are traditional or oriented for Big Data environments, is the development of ETL processes. As discussed in the previous chapters, we do not use this terminology, to avoid confusing ETL and ELT processes, which can cause several unnecessary discussions. Thus, this approach prefers the friendlier NIST terminology (collection and preparation), extending it with the term "enrichment", due to the relevance of derived attributes (feature engineering) for more impactful and actionable insights. As previously discussed, in the proposed approach, these processes are known as CPE processes/workloads. This chapter presents various examples of CPE workloads that practitioners could find useful when implementing BDWs. In these examples, Spark and Talend Open Studio for Big Data are used. Designing and developing CPE workloads for BDWs is arguably one of the most time-consuming and difficult tasks in BDWing. Therefore, structuring various examples that demonstrate typical tasks in these environments is a relevant contribution for the community of practitioners.

7.1. FROM TRADITIONAL DATA WAREHOUSES

Migrations from traditional DWs to BDWs will typically become more common (Russom, 2016). In BDWing implementations, one of the potential workloads will be the migration of the organization's current relational DW to a BDW. This section presents how this task can be achieved using Sqoop, HDFS, Spark, and Hive, four technologies depicted in Figure 7-1. The guidelines here provided are also useful for CPE workloads that read data from relational OLTP databases to fuel the BDW. Sqoop is used to transfer data from relational databases to HDFS, and Spark is used to prepare and enrich the data before storing it in the batch storage component (Hive in this example).

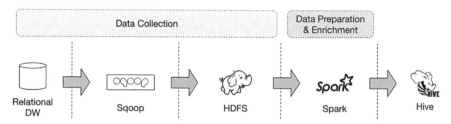

Figure 7-1. CPE workload for traditional DW migration.

Figure 7-1 illustrates the data flow between the components of a possible technological infrastructure. The first step in this process consists in transferring the data from the RDBMS that currently supports the DW to HDFS. This task can be done using Sqoop's import functionality:

```
sqoop import --connect <db_connection_string> {authentication_details} --table
<table_name> --target-dir <path_to_data_folder>
```

In this example, the data from a traditional sales DW modeled according to the SSB benchmark is used, containing one fact table ("`lineorder`") and four dimensions: "`customer`", "`supplier`", "`part`", and "`date`" (O'Neil et al., 2009). After transferring the data and storing it in the distributed file system (HDFS), we can start preparing and enriching this data according to the desired analytical object. In this case, the analytical object is a fully denormalized structure containing all the resulting attributes from the join between the fact table and each dimension, despite the fact that, as seen in section 5.4, analytical objects are flexible and efficient structures that can be more than just a full denormalization of fact tables. The following Spark 2 Python code illustrates a typical script to perform this task:

1. Import Spark packages and classes.

```
from pyspark.sql import SparkSession, Row
from pyspark.sql.types import *
```

2. Define two variables: "`hdfsPath`" and "`hiveDbName`".

```
hdfsPath = "hdfs://<servername>:8020/<path_to_data_folder>/"
hiveDbName = "ssb"
```

3. Create Spark session.

```
spark = SparkSession \
       .builder \
       .appName("Create SSB Analytical Object") \
       .config("spark.sql.warehouse.dir", "/apps/hive/warehouse/") \
       .enableHiveSupport() \
       .getOrCreate()
```

4. Create the Hive database for the BDW.

```
spark.sql("DROP DATABASE IF EXISTS " + hiveDbName + " CASCADE")
spark.sql("CREATE DATABASE " + hiveDbName)
```

5. Create a Spark DataFrame and a Spark Temporary View for each table imported from Sqoop. This will allow the execution of SQL-based instructions on top of the data that has been stored on HDFS.

```
...
dfSchema = StructType([
          StructField("custkey", IntegerType(), True),
          StructField("name", StringType(), True),
          StructField("address", StringType(), True),
          StructField("city", StringType(), True),
          StructField("nation", StringType(), True),
          StructField("region", StringType(), True),
          StructField("phone", StringType(), True),
          StructField("mktsegment", StringType(), True)])
customerDF = spark.read \
          .csv(hdfsPath + "customer", header=False, schema=dfSchema)
customerDF.createGlobalTempView("customer")
...
```

6. Create the Hive table to store the analytical object. In this example, the table uses the ORC file format, which is an optimized columnar format that considerably improves Hive's performance (Huai et al., 2014). The Parquet file format can also be used for Hive tables to achieve adequate performance (Parquet, 2018).

```
spark.sql("CREATE TABLE ssb.analytical_obj (c_custkey int, c_name
varchar(25), ...) STORED AS ORC")
```

© FCA

7. Join the five tables (one fact table and four dimensions) and store the result in the previously created Hive table. If the Hive table is partitioned, the insert statement should reflect the partition scheme, and the adequate HiveQL constructs should be used. This example illustrates a table without partitions.

```
spark.sql("INSERT INTO ssb.analytical_obj SELECT ... FROM
global_temp.lineorder LEFT OUTER JOIN global_temp.customer ON ...")
```

8. Depending on the total size of the resulting table and the number of partitions in the DataFrame, Spark can generate several small ORC files, which can interfere with the performance and adequate operation of Hive and Hadoop. Consequently, the following Hive DDL statement may be useful to concatenate these small ORC files into larger ones. Practitioners may consider this statement in their CPE workloads. Note: there are other ways of manipulating the number of output files, including some Spark configurations and functions (e.g., coalesce and repartition).

```
ALTER TABLE ssb.analytical_obj CONCATENATE
```

9. Finally, it is relevant to highlight the need to ensure that after every CPE workload, the table and column statistics in Hive are adequately computed and refreshed, taking the maximum advantage of this query optimization mechanism. The following Hive DDL is also significantly relevant in these scenarios.

```
ANALYZE TABLE ssb.analytical_obj COMPUTE STATISTICS
ANALYZE TABLE ssb.analytical_obj COMPUTE STATISTICS FOR COLUMNS
```

7.2. FROM OLTP NOSQL DATABASES

In Big Data environments, NoSQL databases are typically the main driver for OLTP workloads, ensuring adequate scalability in intensive random access scenarios (Cattell, 2011). Organizations are currently using NoSQL databases for several applications, for example: massive online sales services (e.g., Amazon); IoT applications; search engines (e.g., Google); and mobile applications.

This section presents a workload for collecting, preparing and enriching data from Cassandra, which is used to store millions of sensor records. Sensors send a record to Cassandra every 15 minutes, which include the following attributes: "`sensor id`"; "`date`" – a timestamp containing the date and time of the record; "`building id`" – the building in which the sensor is located; and "`kwh`" – the energy consumption recorded at that moment. The goal of this workload is to collect Cassandra's data for a specific month and store it in the BDW's batch storage (Hive). The Hive analytical object used for this purpose will be partitioned by year and month. Throughout the workload, the data will be aggregated to match an hourly aggregation level, instead of the original "quarter of an hour" aggregation level. This workload can be written using the following Spark Java code:

1. Import Spark packages and classes. In this example, we use the DataStax open source Spark Cassandra connector.

```
import org.apache.spark.sql.Dataset;
import org.apache.spark.sql.Row;
import org.apache.spark.sql.SaveMode;
import org.apache.spark.sql.SparkSession;
import static org.apache.spark.sql.functions.col;
...
```

2. Create the main Java class and method to include the tasks for this workload. As is already known, the first task defines the Spark Session.

```
SparkSession spark = SparkSession
    .builder()
    .appName("Read sensor records from Cassandra")
    .config("spark.cassandra.connection.host", <hostname(s)>)
    .config("spark.cassandra.auth.username", <username>)
    .config("spark.cassandra.auth.password", <password>)
    .config("spark.sql.warehouse.dir", "/apps/hive/warehouse/")
    .config("hive.exec.dynamic.partition.mode", "nonstrict")
    .enableHiveSupport()
    .getOrCreate();
```

3. Create a Spark Dataset that reads data from the Cassandra table. Spark Datasets are an abstraction introduced in Spark 1.6, which combine the benefits of Spark DataFrames (Spark SQL's optimized execution engine) with the benefits of RDDs, namely strong typing and the ability to use powerful lambda functions (Spark, 2017).

```
Dataset<Row> ds = spark.read().format("org.apache.spark.sql.cassandra")
    .option("keyspace", <keyspace_name>)
    .option("table", <table_name>)
    .load();
```

4. Filter the Dataset to select a specific month (January), and aggregate the Dataset to match an hourly aggregation level.

```
Dataset<Row> dsFiltered = ds.filter(month(col("date")).equalTo(1));

Dataset<Row> dsGrouped = dsFiltered
    .groupBy(
        col("sensor_id"),
        date_format(col("date"), "YYYY-MM-DD HH:00:00").as("moment"),
        col("building_id"))
    .sum("kwh");
```

© FCA

5. Store the Dataset into the corresponding Hive table and partition. Since dynamic partitioning is enabled in the Spark Session configurations, Spark will figure out the partitioning scheme automatically. One needs to be aware that when using the method presented below, the columns of the Dataset must be ordered according to the columns of the Hive table, with the partitioning columns being the last ones. Similarly to the previous section, after this task, one can concatenate small files and recompute table and column statistics.

```
dsGrouped.select(
     col("sensor_id"),
     col("moment"),
     col("building_id"),
     col("sum(kwh)"),
     year(col("moment")),
     month(col("moment")))
  .write().mode(SaveMode.Overwrite).insertInto(<hive_database.table>);
```

7.3. FROM SEMI-STRUCTURED DATA SOURCES

The variety of data is one of the major characteristics for defining Big Data. As was previously highlighted, data may be more or less structured depending on the underlying source. The previous CPE workloads focused on relatively structured data, namely relational and column-oriented schemas. This section focuses on semi-structured data sources, which typically produce data in formats that are not completely detached from a schema, but are flexible or nested, such as server logs, JSON, or XML files.

Take as an example the following GeoJSON file, which is basically a JSON file that, among other attributes, holds geospatial information about buildings in Lisbon:

```
"features": [
   {"type": "Feature",
    "properties": {
       "Shape_Leng": 68.663877,
       "Shape_Area": 276.535056,
       "L_HtRf": 21,
       "Building_Occupation": 3, ...
    },
    "geometry": {
       "type": "MultiPolygon",
       "coordinates": [ [ [
          [ -9.095283006673773, 38.75460513863176, 0.0 ],
          [ -9.095298222128497, 38.754405797462653, 0.0 ], ...] ] ]
    }
   }, ...
```

In the proposed approach, we highlight the use of analytical objects that can contain nested structures. Extracting useful attributes for analysis and implementing an analytical object that adequately handles semi-structured data is key in this specific scenario. In order to handle semi-structured data, one needs to ensure two relevant aspects: i) that the technology used to implement the CPE workload can process these data structures; and ii) that the technology for storing the results of the workload can handle semi-structured data. Regarding the first point, in this section, Talend Open Studio for Big Data is used to build the CPE workload. However, there are many other suitable technologies for this purpose (including Spark). Regarding the second point, Hive is used again as the batch storage for the BDW, since it can adequately handle flexible and nested data structures such as arrays and maps, providing not only ways to store them, but also ways to query and perform analytics on them.

Figure 7-2 illustrates a Talend job used to collect the aforementioned GeoJSON file from HDFS, preparing and enriching it with supplementary GeoJSON files. This job is responsible for fueling a previously created analytical object storing several buildings indicators in Lisbon, including, for example, geospatial information, associated services (e.g., gyms and restaurants), occupation, and construction characteristics.

This job starts by reading the content of the GeoJSON file. Talend Open Studio for Big Data automatically deduces its schema by inspecting a sample of the records within the file. Then, one is able to join the buildings file with supplementary files, such as parishes, neighborhoods (subsections), and services within the building or in its surroundings. The service list can be extracted from the Google Maps Application Programming Interface (API) and, afterwards, one can use several Talend components (e.g., list aggregations and custom java code) to create a list of services associated with each building and make it available in the appropriate format.

Finally, after all the previous tasks are completed, the data is sent to HDFS and a temporary Hive table is created to store the data in text format. As previously noted, Hive tables in ORC or Parquet format are more suitable for analytical purposes, so one needs to move the data from this temporary table into a table using the ORC format. This procedure involving the use of a temporary table is common in Hive-based DWs. However, practitioners may find other ways to directly move the data to ORC tables without a temporary table, for example, using the ORC API. The final result is an analytical object stored as a Hive table, which can handle a variety of structures, including arrays and maps.

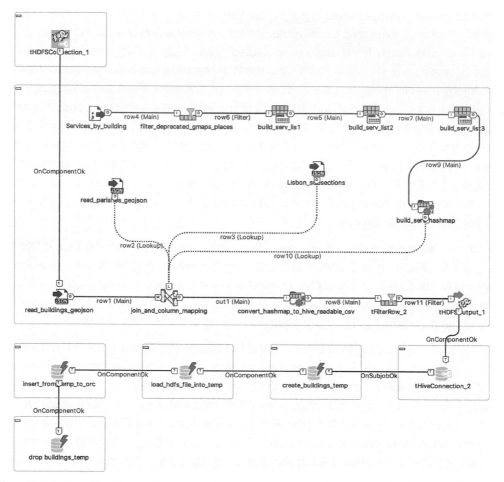

Figure 7-2. CPE workload for semi-structured data.

7.4. FROM STREAMING DATA SOURCES

So far, only the use of the BDW's batch storage component was demonstrated. When a source generates data through streaming mechanisms, one needs to rely on the BDW's streaming storage. Data velocity is another relevant characteristic of Big Data environments and, in this section, we will discuss the development of a streaming CPE workload to fuel the BDW. Kafka is used for data collection, Spark Streaming is used for data preparation and enrichment, and Cassandra is used as the NoSQL database responsible for the BDW's streaming storage. Figure 7-3 summarizes this CPE workload.

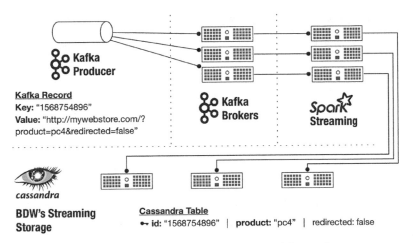

Figure 7-3. Streaming CPE workload using Kafka, Spark Streaming, and Cassandra.

1. The first step is to develop a Kafka producer. In this example, the producer generates a record every five seconds, corresponding to a random product sale in a simulated e-commerce environment. Each record contains a key ("`sales id`") and a value ("`URL`" of the webpage where the product was purchased). The following Java code snippets demonstrate this scenario, and can be used as a guide for other Kafka producers.

   a) Import the Java packages and classes. For this producer, the Apache Kafka API is used.

```
...
import org.apache.kafka.clients.producer.KafkaProducer;
import org.apache.kafka.clients.producer.ProducerRecord;
import org.apache.kafka.clients.CommonClientConfigs;
import org.apache.kafka.clients.producer.Callback;
import org.apache.kafka.clients.producer.ProducerConfig;
import org.apache.kafka.clients.producer.RecordMetadata;
```

   b) Create the Java class and its variables. Generally, this class needs the topic in which the producer will publish the records, the producer object, and several properties that reflect the infrastructure in use and the application requirements (e.g., secure/unsecure cluster and the number of acknowledgments to consider a request as being completed). In this example, there is also one variable containing random products used to generate random online sales URLs.

```
public class DummyProducer extends Thread {
    private final String topic;
    private final KafkaProducer<String, String> producer;
    private final Properties props;
    private final String[] products;
...
```

c) Create the constructor that initializes the variables mentioned above.

```
public DummyProducer(String topic, String kafkaServerUrl, int
kafkaServerPort)
{
   this.products = new String[] {
      "smartphonex7", "pc4", "keyboardy", "monitorpro"
   };

   this.props = new Properties();
   this.props.put(ProducerConfig.BOOTSTRAP_SERVERS_CONFIG,
      kafkaServerUrl + ":" + kafkaServerPort);
   this.props.put(ProducerConfig.CLIENT_ID_CONFIG, "DummyProducer");
   this.props.put(CommonClientConfigs.SECURITY_PROTOCOL_CONFIG,
      "SASL_PLAINTEXT");
   this.props.put(ProducerConfig.ACKS_CONFIG, "all");
   this.props.put(ProducerConfig.KEY_SERIALIZER_CLASS_CONFIG,
      "org.apache.kafka.common.serialization.StringSerializer");
   this.props.put(ProducerConfig.VALUE_SERIALIZER_CLASS_CONFIG,
      "org.apache.kafka.common.serialization.StringSerializer");

   this.producer = new KafkaProducer<>(props);
   this.topic = topic;
}
```

d) Create the run method responsible for executing the main task, i.e., generating an URL for a random online sale every five seconds for an infinite timespan. In this example, a random product is selected from a set of four available products (see code snippet above). Each sale also contains a random flag indicating whether the sale was the result of a recommendation based on a previous visualized product ("`redirected`" attribute).

```
@Override
public void run() {
   Random random = new Random();
   String message;

   while(true) {
      String salesID = "" + random.nextInt() + System.currentTimeMillis();
      message = String.format(
         "\"http://mywebstore.com/?product=%s&redirected=%s\"",
         this.products[random.nextInt(4)], random.nextBoolean());

      ProducerRecord<String, String> data = new ProducerRecord<>(
         this.topic, salesID, message);
      this.producer.send(data);
      try {
         Thread.sleep(5000);
```

```
    } catch (InterruptedException ex) {
       System.err.println(ex.getMessage());
    }
  }
}
```

e) Finally, create the main method for the "`DummyProducer`" class, which will simply run the Kafka producer with the given Kafka topic, broker, and port.

```
public static void main(String args[]) {
   DummyProducer producer = new DummyProducer(<topic>, <kafka_broker>, <port>);
   producer.start();
}
```

2. Having a streaming producer is just part of the CPE workload, specifically it represents the collection stage of the workload. Consequently, in BDWing environments, one typically needs to prepare and enrich the data before making it available for analytical purposes. One way to achieve this goal is using the powerful and stable Spark Streaming API, which allows to use multiple functions (e.g., filter, join, count, and map) on streaming sources like Kafka, ensuring adequate scalability and fault-tolerance. The following Java code snippets demonstrate a Spark Streaming application that uses regular expressions to extract information from the Kafka messages and storing the results in the BDW's streaming storage component. In this example, Cassandra is used to store a streaming analytical object containing the "`sales id`", the "`product`", and the "`redirected`" attributes.

a) Import the required packages for this Spark Streaming application. The key APIs are the Spark Core API, the Spark Streaming API, the Spark Streaming Kafka API, and the DataStax Cassandra connector.

```
...
import static com.datastax.spark.connector.japi.CassandraJavaUtil.javaFunctions;
import static com.datastax.spark.connector.japi.CassandraJavaUtil.mapToRow;
import org.apache.spark.SparkConf;
import org.apache.spark.streaming.api.java.*;
import org.apache.spark.streaming.kafka010.*;
import org.apache.kafka.clients.consumer.ConsumerRecord;
import org.apache.kafka.common.serialization.StringDeserializer;
import org.apache.spark.api.java.JavaRDD;
import org.apache.spark.streaming.Durations;
```

b) After creating the main class and the main method for the Spark Streaming application, configure the Kafka connector, including the consumer configurations and the list of topics to be consumed. The following code snippet illustrates the configuration of a Kerberized cluster, in which the Spark consumer informs Kafka when it has finished consuming a certain offset; that is why one disables Kafka auto commits and uses the "`offsetRanges`" variable. This ensures that the

Spark consumer only acknowledges the processing of certain offsets when the records are already stored in Cassandra. We will clarify this functionality later in this subsection.

```
Map<String, Object> kafkaParams = new HashMap<>();
    kafkaParams.put("bootstrap.servers", <kafka_broker>:<port>);
    kafkaParams.put("key.deserializer", StringDeserializer.class);
    kafkaParams.put("value.deserializer", StringDeserializer.class);
    kafkaParams.put("group.id", "spark.events");
    kafkaParams.put("auto.offset.reset", "latest");
    kafkaParams.put("enable.auto.commit", false);
    kafkaParams.put("security.protocol", "SASL_PLAINTEXT");

    Collection<String> topics = Arrays.asList(<topic(s)>);
    final AtomicReference<OffsetRange[]> offsetRanges = new AtomicReference<>();
```

c) Create the Spark configuration and the Spark Streaming object. Since we are storing the results in Cassandra, the Cassandra connection properties are also needed, just like in the previous OLTP NoSQL-based CPE workload. In this example, the streaming application processes data arriving from Kafka in 10 seconds micro batch intervals. As noted in subsection 5.2.3.2, micro batches are configurable, and they are often a trade-off between latency, throughput, and flexibility.

```
SparkConf conf = new SparkConf()
    .setAppName("StreamingCPEWorkload")
    .set("spark.cassandra.connection.host", <host(s)>)
    .set("spark.cassandra.auth.username", <username>)
    .set("spark.cassandra.auth.password", <password>);

JavaStreamingContext jssc = new JavaStreamingContext(
    conf, Durations.seconds(10));

JavaInputDStream<ConsumerRecord<Integer, String>> stream =
    KafkaUtils.createDirectStream(
        jssc,
        LocationStrategies.PreferConsistent(),
        ConsumerStrategies.Subscribe(topics, kafkaParams));
```

d) As previously stated, this application is using manual Kafka commits and, therefore, one needs to inform Kafka when the data has been processed. Consequently, the first step after creating the stream is to store the Kafka offset range in the Spark application, to commit the processed offsets after the data has been successfully stored in Cassandra. The following function needs to be the first one called after creating the stream, since it does not work once the transformations to the "stream" object have been applied. This ensures the application has "exactly-once" semantics instead of "at-least-once" semantics, which makes sure the data arriving from Kafka is not processed and stored twice.

```
stream.foreachRDD((JavaRDD<ConsumerRecord<Integer, String>> rdd) -> {
    OffsetRange[] offsets = ((HasOffsetRanges) rdd.rdd()).offsetRanges();
    offsetRanges.set(offsets);
});
```

e) To extract the "`product`" and "`redirected`" attributes arriving from Kafka's messages, one can use regular expressions on the URL. The "`map`" transformation can be used to extract these attributes. Spark Streaming offers several transformations, window, join, and output functions that can be used for streaming contexts. For example, joining a stream with an historical dataset can be significantly useful for BDWing purposes, to prepare and enrich data for certain analytical objects.

```
JavaDStream<DummySale> transformedStream = stream
    .map((ConsumerRecord<Integer, String> event) -> {
        String[] fields = event.value().split("\";\"");
        Matcher m = Pattern.compile("product=(.*)&redirected=(.*)").matcher(fields[1]);
        m.find();
        return new DummySale(
            fields[0], m.group(1), Boolean.parseBoolean(m.group(2)));
    });
```

f) Once all the processing tasks in this example are completed, the results can be stored in Cassandra, and the Spark Streaming application can then commit the offsets to Kafka, acknowledging that the specific records have been processed and stored. It should be noted that, in this example, "`DummySale`" is a typical Java Bean containing the same attributes as the analytical object stored in Cassandra. This Java Bean is used to apply a schema to each row in the Spark RDD.

```
transformedStream.foreachRDD((JavaRDD<DummySale> rdd) -> {
    javaFunctions(rdd).writerBuilder(
        <topic>, <cassandra_database>, mapToRow(DummySale.class)
    ).saveToCassandra();
    ((CanCommitOffsets) stream.inputDStream()).commitAsync(offsetRanges.get());
});
```

g) Finally, the last task consists in simply starting the application and waiting for its termination.

```
try {
    jssc.start();
    jssc.awaitTermination();
} catch (InterruptedException ex) {
    System.err.printf("The application '%s' has stopped! ", conf.getAppId());
}
```

© FCA

7.5. USING DATA SCIENCE MODELS

One of the main aspects in the proposed approach (previously described in Chapter 5) is the inclusion of data science models in CPE workloads fueling the BDW. This book considers data science as an umbrella term for several related and more specific subareas, including: data mining/machine learning; text mining; image mining; and video mining. Regarding data mining, traditional DWs are frequently considered a relevant data source for the algorithms used in this area, since they typically contain an extensive record of historical data regarding the organization. Since these algorithms need a vast training set to extract patterns, traditional DWs are natural sources of data for feeding these algorithms, which can be considered "clients of the DW" (Kimball & Ross, 2013). This is also true for a BDW (Figure 5-2), wherein it can be queried by data scientists that are "playing" with the data in the data science sandbox. However, this book extends this idea by inviting practitioners to include data mining/machine learning algorithms in CPE workloads, in order to create new predictive attributes and include them in the analytical objects stored in the BDW (Figure 5-3).

The same strategy remains valid for unstructured data science techniques like text mining, image mining, or video mining. These techniques are not frequently seen in traditional DWing environments. One could argue that raw unstructured data has almost no value for analytical purposes. Patterns should be extracted using data science techniques and then, since these results are already relatively structured, they can follow their path to an OLAP-oriented system like the BDW. Another argument that can be made is that the rigid structure of relational DWs is a significant barrier in these scenarios, since most of the time it is unnatural, time-consuming, and inefficient to model dimensions, fact tables, and relationships for this type of analytical workloads.

For example, when an organization is collecting images in real-time and instantaneously using an algorithm to predict the occurrence of a certain pattern in that image (e.g., template matching for manufacturing quality control), it becomes really inefficient to fuel a relational DW via streaming mechanisms. For each image being analyzed, a typical ETL process has to scan the several dimensions for any changes since the last DW refresh (e.g., new rows to add/update in dimension tables), or to retrieve each dimension's SK for matching the FK of each new row in the fact table, for example. In these contexts, relational DWs are not the most adequate solution, and fully denormalized structures (analytical objects), are arguably more efficient and simpler to implement, since their corresponding CPE workloads are considerably easier to develop and maintain compared to traditional ETL processes.

In Big Data environments, there is the need to integrate both structured and unstructured sources (Kimball & Ross, 2013). As previously discussed, predictive analytics is also a relevant use case that BDWs must consider among their set of mixed and complex analytical workloads. This is the reason why including structured and unstructured data science models in CPE workloads can be seen as a way of extracting value hidden in Big Data, which can then be used to make predictions about future events and to fuel the analytical objects stored in the BDW. This section discusses two types of CPE workloads including data science models, using data mining/machine learning models for structured data and using text mining, image mining, and video mining models for unstructured data.

7.5.1. Data Mining/Machine Learning Models for Structured Data

Predictive attributes are the key for predictive analytics within BDWs. We focus on the use of data science models to create these attributes. The data stored either in the file system or in the indexed storage of the BDW can be used to train these predictive models, which can then be used to enrich data arriving at the system with new predictive attributes. In these contexts, data mining/machine learning models can be useful for CPE workloads dealing with structured data. This subsection uses the Spark MLlib API to demonstrate one of many data mining techniques that can be used in BDWing systems, namely clustering.

Clustering can be used when the training dataset is not previously labeled with the attribute one wants to predict. In this case, this is also mentioned as an unsupervised learning task. There are many other techniques available in Spark MLlib, both unsupervised (e.g., association rules) or supervised (e.g., classification and regression), offering scalable ways to train, test, and apply data mining/machine learning models. The following Java code snippets demonstrate the use of clustering algorithms in Spark, namely the very broadly used K-means algorithm. Figure 7-4 presents an overview of the CPE workload implemented in this subsection.

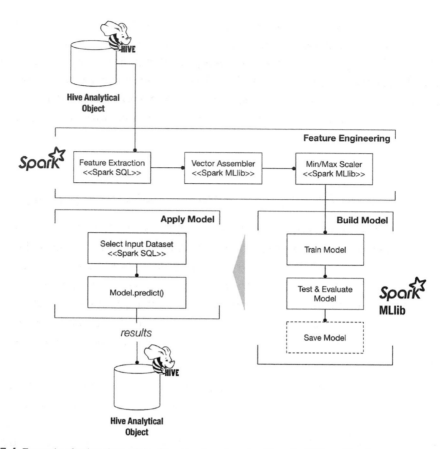

Figure 7-4. Example of using data mining/machine learning algorithms in CPE workloads.

1. Import the java packages needed for the application.

```
...
import org.apache.spark.sql.SparkSession;
import org.apache.spark.ml.clustering.KMeansModel;
import org.apache.spark.ml.clustering.KMeans;
import org.apache.spark.ml.feature.MinMaxScaler;
import org.apache.spark.ml.feature.MinMaxScalerModel;
import org.apache.spark.ml.feature.VectorAssembler;
import org.apache.spark.ml.linalg.Vector;
import org.apache.spark.sql.Dataset;
import org.apache.spark.sql.Encoders;
import org.apache.spark.sql.Row;
```

2. Create the main class and the main method, which starts by initiating the Spark Session. In this example, one will be segmenting customers according to their buying behavior (following the example of section 7.1): how many orders do they place? How much revenue do they bring to the company? Are they regular monthly customers?

```
SparkSession spark = SparkSession
   .builder()
   .appName("Segmenting Customers using K-means")
   .config("spark.sql.warehouse.dir", "/apps/hive/warehouse/")
   .config("hive.exec.dynamic.partition.mode", "nonstrict")
   .enableHiveSupport()
   .getOrCreate();
```

3. To accomplish this goal, the analytical object created in subsection 7.1 can be used, which, as already seen, is based on the SSB dataset (O'Neil et al., 2009). Since the analytical object corresponds to data that is already stored in the BDW, this workload does not include a data collection stage from an external data provider. This is a workload that uses the models, insights, and results derived from the data science sandbox component of the BDW and, therefore, the data science sandbox can be regarded as a data provider. In this example, using Spark SQL, one can submit a query to the Hive batch storage component, to retrieve the training set needed to segment customers.

```
Dataset<Row> customerSales = spark.sql("
WITH

customerSales AS (
SELECT c_custkey, c_name, c_city, c_nation, c_region, c_mktsegment,
od_monthnuminyear, sum(quantity) as monthly_quantity, sum(revenue) as
monthly_revenue, count(1) as monthly_orders
FROM ssb_analytical_objects.analytical_obj10
GROUP BY c_custkey, c_name, c_city, c_nation, c_region, c_mktsegment,
od_monthnuminyear
),

minmax AS (
SELECT c_custkey, MIN(monthly_revenue) min_monthly_revenue,
MAX(monthly_revenue) max_monthly_revenue
FROM customerSales
GROUP BY c_custkey)

SELECT customerSales.c_custkey, c_name, c_city, c_nation, c_region,
c_mktsegment, stddev((monthly_revenue -
min_monthly_revenue)/(max_monthly_revenue - min_monthly_revenue))
revenue_monthly_stddev, sum(monthly_revenue) revenue, sum(monthly_orders)
total_orders, sum(monthly_quantity) quantity
FROM customerSales
LEFT OUTER JOIN minmax ON customerSales.c_custkey = minmax.c_custkey
GROUP BY customerSales.c_custkey, c_name, c_city, c_nation, c_region,
c_mktsegment
");
```

© FCA

4. One of the main tasks in data mining/machine learning processes is feature engineering. In this example, the previous query created a training set with customers and their respective orders, revenue, and standard deviation regarding monthly revenue. However, one way to improve the efficiency of learning algorithms is called feature scaling, i.e., providing a standard scale for all features. Moreover, in Spark MLlib 2, all features must be contained in a `Vector` object and, therefore, one can use the `VectorAssembler` to transform the original data from Hive, while at the same time replacing null values with zeros, so that the `VectorAssembler` can be properly used. The following code snippet illustrates these simple feature engineering tasks. Nonetheless, Spark offers several other functions for these purposes.

```
VectorAssembler assembler = new VectorAssembler()
    .setInputCols(new String[]{"revenue_monthly_stddev", "revenue", "total_orders"})
    .setOutputCol("vectors");
Dataset<Row> vectorizedData = assembler.transform(customerSales.na().fill(0));
MinMaxScaler scaler = new MinMaxScaler()
    .setInputCol("vectors")
    .setOutputCol("features");
MinMaxScalerModel scalerModel = scaler.fit(vectorizedData);
```

5. The next step consists in training and testing the K-means model with the previously prepared features. After training the model, data scientists can evaluate its performance, by changing the number of clusters that will be created and analyzing the behavior of the cluster sum of squared errors, and by manually inspecting the clusters' centers, for example. This evaluation allows to understand what each cluster means. For example: one cluster may represent customers with less orders, but who bring more revenue for the company in steady monthly rates; in contrast, another cluster may represent irregular customers with several orders, but bringing low income for the company. In this workload, for demonstration purposes, the selected number of clusters is 2.

```
KMeans kmeans = new KMeans().setK(2).setSeed(1L);
KMeansModel model = kmeans.fit(trainingSet);

double wssse = model.computeCost(trainingSet);
System.out.println("Within Set Sum of Squared Errors = " + wssse);

Vector[] centers = model.clusterCenters();
System.out.println("Clusters Centers: ");
for (Vector center : centers) {
    System.out.println(center);
}
```

6. Despite the fact that in this workload we trained, tested, and applied the model in the same Spark application, both the feature engineering models and the K-means model can be permanently saved and used for later executions of this or other workloads. This means that the models do not need to be trained each time that they are applied.

```
try {
  model.write().save("<path_in_hdfs>");
} catch (IOException ex) {
  System.err.println("Error saving the model!");
}
```

7. The last step is storing the results into the new Hive analytical object. The inclusion of data science models in CPE workloads does not always involve generating new analytical objects. Sometimes, the workload simply refreshes existing analytical objects with new data. On other ocassions, existing analytical objects can be updated with new attributes (e.g., Hive supports different schemas for different partitions in a table). In this workload, since one is training, testing, and applying the learning model using a single application, each time the workload is executed the analytical object is created/overwritten. This analytical object contains attributes similar to those in the "`customerSales`" Spark Dataset used to train the K-means model, but with the addition of the cluster to which the customer belongs, along with a user-friendly description of the cluster according to its centroid. This is possible by applying the predict method from the K-means model. Having an analytical object containing the customers' buying behavior allows to carry out interesting analyzes, even allowing the join between this analytical object and the original one containing all sales transactions. It should be noted that, in this example, "`customerSale`" is a typical Java Bean containing the same attributes as the analytical object stored in Hive.

```
Dataset<CustomerSale> analyticalObject = trainingSet.map((Row r) -> {
    int cluster = model.predict((Vector) r.get(11));
    String levConstantIncome;
    String levRevenueGenerated;
    String levTotalOrders;
    switch (cluster) {
        case 0:
            levConstantIncome = "Buys more frequently";
            levRevenueGenerated = "Low";
            levTotalOrders = "Low";
            break;
        case 1:
            levConstantIncome = "Buys less frequently";
            levRevenueGenerated = "Average-Higher";
            levTotalOrders = "Average-Higher";
            break;
        default:
            levConstantIncome = "NA";
            levRevenueGenerated = "NA";
            levTotalOrders = "NA";
            break;
    }
```

```
    CustomerSale c = new CustomerSale();
    c.setC_custkey(r.getInt(0));
    c.setC_name(r.getString(1));
    c.setC_city(r.getString(2));
    c.setC_nation(r.getString(3));
    c.setC_region(r.getString(4));
    c.setC_mktsegment(r.getString(5));
    c.setRevenue_monthly_stddev((int) r.getDouble(6));
    c.setRevenue((int) r.getDouble(7));
    c.setTotal_orders(r.getLong(8));
    c.setCluster("cluster" + cluster);
    c.setLev_constant_income(levConstantIncome);
    c.setLev_revenue_generated(levRevenueGenerated);
    c.setLev_total_orders(levTotalOrders);
    return c;
}, Encoders.bean(CustomerSale.class));

analyticalObject.write().mode(SaveMode.Overwrite).insertInto("<hive table>");
```

7.5.2. Text Mining, Image Mining, and Video Mining Models

Although unstructured data mining is relatively different from structured data mining, the general steps presented in the previous subsection can still be applied. For this reason, as Figure 5-3 demonstrates, the proposed method for CPE is fairly similar for both structured and unstructured data. Obviously, while one can use classification, regression, clustering, association rules, or time series forecasting to extract patterns and make predictions in structured environments (Pujari, 2001), in unstructured contexts, the techniques may be significantly different (although sometimes they overlap). Regarding the technologies to be used in these contexts, they depend on the specific use case. For example, Spark MLlib does not have an extensive set of text mining algorithms, but it offers some text-based feature extraction and clustering algorithms (e.g., TF-IDF, Word2Vec, and LDA). However, currently, Spark does not offer adequate support for image or video mining algorithms. In these contexts, choosing complementary technologies (such as Python, which offers some interesting libraries oriented towards image mining) in the data science sandbox is appropriate.

In BDWing environments, including unstructured data science models in CPE workloads is intended to extract structured predictive attributes, which are structured findings extracted from unstructured sources. These attributes can be considered the structured value that can be extracted from unstructured data, which by itself, in its raw state, would not be significantly relevant for BDWing purposes. The data has to be prepared and enriched using adequate techniques and technologies capable of mining the value from these sources. Only then the results of these tasks will provide analytical value.

Figure 7-5 presents a workflow based on Figure 5-3, including several techniques useful in these scenarios. As can be seen, despite the challenges and complexity of unstructured data mining, the general tasks remain similar to a CPE workload that includes data mining/ machine learning algorithms for structured data. First, the data is collected using batch or streaming mechanisms. For a specific source and a specific technique, a previously trained model is used to extract structured patterns from text, images, or video, depending on the use case. Complementary datasets can also be used for data enrichment, if applicable. After all the (descriptive, factual, and predictive) attributes of the analytical object are created, the analytical object is ready to be used. So far, there is no difference compared to the CPE workload discussed in the previous subsection.

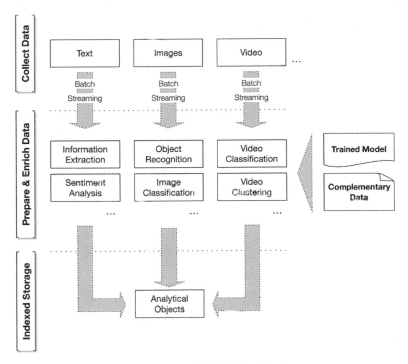

Figure 7-5. Including unstructured data science models in CPE workloads.

That being said, the difference relies solely on the use of new and challenging techniques: for text mining, techniques such as information extraction and sentiment analysis can be significantly useful for extracting entities (e.g., people and dates), relationships (e.g., events), and sentiments from raw text (Gandomi & Haider, 2015). This will allow fueling analytical objects which can be significantly useful for many organizations; for image mining purposes, techniques such as object recognition (e.g., template matching) and image classification can also be significantly useful (Zhang, Hsu, & Lee, 2001); finally, for video mining, video classification and video clustering can be used (Vijayakumar

& Nedunchezhian, 2012), which have similar goals as their structured data mining counterparts, although, of course, with different specifications. These are just some examples of possible techniques, since the list can be considerably extended. However, to demonstrate their role in BDWing environments, these techniques provide adequate examples of the capabilities of unstructured analytics in BDWs.

8 EVALUATING THE PERFORMANCE OF BIG DATA WAREHOUSES

INTRODUCTION

This chapter discusses the evaluation of BDWs built using the proposed approach. To evaluate the performance of a BDW, several related benchmarks can be used, such as the TPC-DS benchmark (TPC, 2017a) or the SSB benchmark (O'Neil et al., 2009), for example. In this work, an extension of the SSB benchmark, named SSB+ (C. Costa & Santos, 2018), which combines batch and streaming data, was specifically created for BDWing contexts. An extension of the original SSB benchmark was needed due to the lack of workloads that combine volume, variety, and velocity of data, with adequate customization capabilities and integration with current versions of different Big Data technologies. Moreover, one needs to evaluate different modeling strategies (e.g., flat structures, nested structures, and star schemas) and different workload considerations (e.g., partitioned analytical objects and dimensions' size in star schema-based BDWs) and, therefore, an adaptation of the SSB benchmark is required. This chapter presents the SSB+ Benchmark, discussing the performance, advantages, and disadvantages of several design and implementation choices in the proposed approach, extending and integrating previously published scientific works (C. Costa & Santos, 2018; E. Costa, Costa, & Santos, 2017).

8.1. THE SSB+ BENCHMARK

This section details the SSB+ Benchmark, namely the data model, queries, system architecture, and infrastructure. Besides serving as a proof-of-concept validation, presenting several insights related to relevant design decisions for BDWs, the SSB+ Benchmark is useful for practitioners who want to evaluate the performance of their own implementations.

8.1.1. Data Model and Queries

The SSB+ Benchmark data model (C. Costa & Santos, 2018), presented in Figure 8-1, is based on the original SSB benchmark (O'Neil et al., 2009), so all the original tables remain the same ("lineorder", "part", "supplier", and "customer"), with the exception of the "date" dimension, which has been streamlined to remove the temporal attributes that are not used in the 13 original SSB queries. These 13 queries were not modified besides the replacement of "where clause" joins for ANSI SQL joins with an explicit join operator. This measure is taken to ensure an optimal execution plan for the optimizers of the query engines.

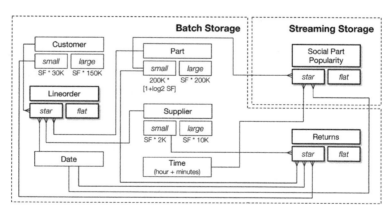

Figure 8-1. SSB+ data model. Adapted from O'Neil et al. (2009) and C. Costa & Santos (2018) with extended content.

Since the original SSB benchmark only takes into consideration a star schema-based DW, the SSB+ also includes jobs for transforming the "lineorder" star into a flat "lineorder" analytical object. The original 13 SSB queries are also modified to match the new flat analytical object. These changes allow us to compare the advantages and disadvantages of star schemas and flat structures for BDWs. Moreover, the SSB+ also considers two different sizes in the dimensions: the original TPC-H sizes (TPC, 2017b) (benchmark in which the original SSB is based), which includes larger "part", "customer", and "`supplier`" tables; and the original SSB sizes, in which these tables are smaller to represent more traditional dimensions in the retail context. This SSB+ feature allow us to understand the impact of the dimensions' size in star schema-based BDWs. Furthermore, the SSB+ also includes a "`returns`" table (flat analytical object and star schema fact table) and four new queries to evaluate the performance of drill across operations, as well as window and analytics functions.

Regarding the streaming workloads of the SSB+ Benchmark, a new "`time`" dimension table is included, as the data stream has a "`minute`" granularity. This new dimension can then be joined with the new "`social part popularity`" fact table, as well as other existing dimensions such as "`part`" and "`date`". A flat version of this fact table is also available for performance comparison purposes. The "`social part popularity`" table contains data from a simulated social network, where users express their sentiments regarding the parts sold by the organization represented in the SSB and SSB+ Benchmark. Along with these new tables, three new streaming queries were developed for both the star schema-based BDW and the flat-based BDW, performing several aggregation, filtering, union, and join operations on streaming data. All the applications, scripts, and queries for the SSB+ Benchmark can be found in (C. Costa, 2017).

8.1.2. System Architecture and Infrastructure

The SSB+ Benchmark takes into consideration various technologies to accomplish different goals, from data CPE workloads to querying and OLAP tasks. These technologies are presented in Figure 8-2. Starting with the CPE workloads, for batch data, the SSB+ considers a Hive script with several beeline commands that load the data from HDFS to the Hive tables stored in the ORC format, an efficient columnar file format for data analytics. Several SFs can be generated using the original SSB generator. This book considers the SF=30, SF=100, and SF=300 for the batch performance evaluation. Regarding streaming data, a Kafka producer generates simulated data at configurable rates, and this data is processed by a Spark Streaming application that finally stores it in Hive and Cassandra. Streaming data is stored both in Hive and Cassandra for benchmarking purposes (see section 8.3).

For querying and OLAP, this book considers both Hive on Tez and Presto, which are two robust and efficient SQL-on-Hadoop engines (Santos et al., 2017). Obviously, practitioners can run the SSB+ Benchmark with any SQL-on-Hadoop engine of their choosing, as long as they develop the adequate scripts to run the workloads. Currently, the repository mentioned in the previous subsection contains only applications and scripts supporting the technologies mentioned in Figure 8-2. However, all the contents of the repository are open to the public to facilitate changes or extensions. Hive and Presto are used to provide

© FCA

insights from different SQL-on-Hadoop engines, to see if the conclusions hold true for more than one engine, since one of them may perform better under certain data modeling strategies. However, in the streaming workloads, only Presto is used, since it targets interactive SQL queries over different data sources, including NoSQL databases, which is not a very proclaimed feature in Hive, although it can also be used for this purpose. Moreover, despite Tez's tremendous improvements to Hive's performance, Hive on Tez may not be considered a low latency engine, as the results presented in this chapter may suggest.

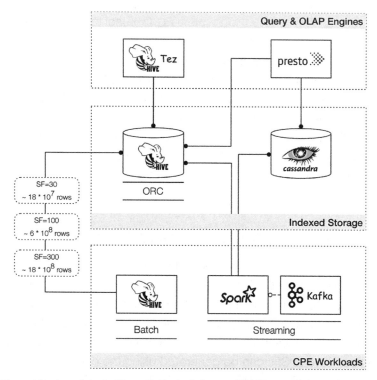

Figure 8-2. SSB+ architecture. Adapted from C. Costa & Santos (2018).

The infrastructure used in this work is a 5-node Hadoop cluster with one HDFS NameNode (YARN ResourceManager) and four HDFS DataNodes (YARN NodeManagers). The hardware used in each node includes:

- 1 Intel Core i5, quad-core, with a clock speed ranging between 3.1GHz and 3.3GHz;

- 32GB of 1333MHz DDR3 RAM, with 24GB available for query processing;

- 1 Samsung 850 EVO 500GB Solid State Drive (SSD), with up to 540MB/s read speed and up to 520MB/s write speed;

- 1 Gigabit Ethernet card connected through Cat5e Ethernet cables and 1 Gigabit Ethernet switch.

The operative system is CentOS 7 with an XFS file system, and the Hadoop distribution is the Hortonworks Data Platform 2.6. Besides Hadoop, a Presto coordinator is also installed on the NameNode, as well as four Presto workers on the four remaining DataNodes. All configurations are left unchanged, apart from the HDFS replication factor, which is set to 2, as well as Presto's memory configuration, which is set to use 24GB of the 32GB available in each worker (identical to the memory available for YARN applications in each NodeManager).

8.2. BATCH OLAP

Batch OLAP queries take as input vast amounts of data stored in the batch storage component of the BDW. This section discusses the performance of batch OLAP queries for BDWs using two modeling approaches: star schemas and flat analytical objects. Moreover, this section also addresses the impact of the dimensions' size on star schemas, the use of nested structures in analytical objects, the improvement of the BDW's performance by using adequate data partitioning, and the performance of drill across queries and window, as well as analytics functions.

8.2.1. Comparing Flat Analytical Objects with Star Schemas

This first evaluation consists in analyzing the performance, storage size, CPU usage, and memory requirements of flat analytical objects and star schemas, using the 13 SSB+ batch queries. Regarding the star schema, all the workloads depicted in this subsection use the larger dimensions instead of the smaller ones, which will be discussed in subsection 8.2.3.

Analyzing the small to medium SFs, illustrated in Figure 8-3, we can conclude that the performance advantage of flat analytical objects is quite noticeable. For the majority of the queries, the flat object visibly outperforms the star schema, especially in the SF=100 workload, wherein the performance of a star schema with a high number of rows starts to degrade. Interestingly, such phenomenon does not hold true for Hive's Q2.2, and the star schema shows better performance in this scenario, possibly due to some performance problems in Hive's ability to process string range comparisons ("p_brand1 between 'MFGR#2221' and 'MFGR#2228'") with significantly larger amounts of data (see Figure 8-4 to understand the storage size impact of the flat analytical object). This phenomenon does not occur in the Presto SQL-on-Hadoop engine.

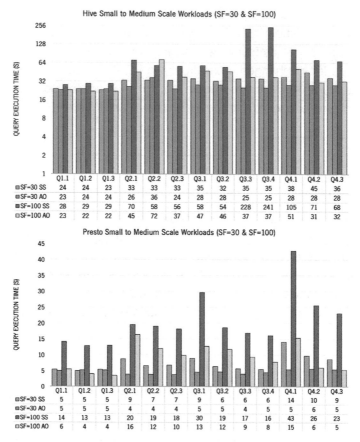

Figure 8-3. Small to medium batch SSB+ workloads.
Star schema (SS); analytical object (AO). Hive's results are based on E. Costa et al. (2017).

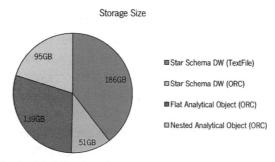

Figure 8-4. Storage size for the SF=300 using different modeling approaches.

Moreover, looking into Hive's `Q1.1`, `Q1.2`, and `Q1.3`, the flat analytical object and the star schema performance are fairly similar, which is comprehensible, since for these queries, the star schema only needs to join the fact table "`lineorder`" with the dimension "`date`". Given that the flat analytical object is around 2.5 times bigger than the corresponding dimensional DW stored in the ORC file format (see Figure 8-4), it balances out the cost of the join operation. However, in Presto's SF=100 workload, despite this fact, the flat analytical object still outperforms the star schema. At this point, Presto started exhibiting a significantly satisfactory performance when using completely flat structures.

Regarding the large-scale batch workload (SF=300), depicted in Figure 8-5, the trend continues, specifically the overall performance advantage of using flat structures. The performance of a Hive star schema for most of the queries is not satisfactory for interactive scenarios, often being more than 3 to 4 times slower than a flat structure. There are some exceptions (Hive/Presto's `Q1.1`, `Q1.2`, and `Q1.3`; and Hive's `Q4.3`), mainly due to the aforementioned reasons, i.e., the storage size of the flat structure causes a significant overhead in the I/O tasks of the queries, which turns them into I/O bound queries, and causes the flat analytical object to perform worse than the star schema. Consequently, even though flat structures tend to perform significantly better than star schemas in these environments, there are certain queries wherein joining a fact table with a small dimension (e.g., "`date`" dimension) is faster than executing the same queries on flat structures. However, one also needs to consider the storage size of these two data sources. Looking at Figure 8-4, the entire star schema DW using the ORC file format has around 51GB, whereas its flat counterpart has around 139GB. Considering the infrastructure used for this book and previously described in subsection 8.1.2, the entire star fits into memory, whereas the flat analytical object significantly surpasses the total amount of memory available for querying.

Smaller dimensions allow an efficient type of join, known as broadcast join (or map join in Hive) (Floratou et al., 2014). When using broadcast joins, the smaller tables involved in the join operation are broadcasted to the memory of the nodes involved in the computation, which means that the large table (traditionally a fact table) is joined with all these structures in memory, while it is being processed throughout the nodes. The effects of using broadcast joins can be seen in Figure 8-5, in which Presto exhibits a significant decrease in query execution times, compared to the more conventional distributed hash join. However, despite this advantage, Presto is even faster when using flat structures that do not require joins. Such results do not favor the dimensional approach for DWs in Big Data environments.

© FCA

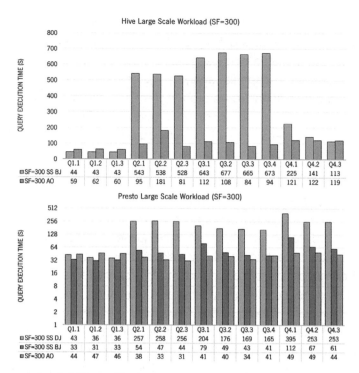

Figure 8-5. Large-scale batch SSB+ workload.
Star schema with broadcast joins (SS BJ); analytical object (AO); star schema with distributed hash joins (SS DJ). Hive's results are based on E. Costa et al. (2017).

Doing broadcast joins is not always possible, since this technique requires that the dimensions fit into a fraction of the memory available for query processing, which is not always the case if the dimensions are naturally large or become larger through the application of type 2 SCD techniques (Jukic et al., 2017; Kimball & Ross, 2013). Certain query optimizers do not automatically select the most appropriate join technique according to the size of the tables, which is the case of Presto's optimizer in version 0.180. For this book, the two join techniques (distributed and broadcast) were manually selected. When enforcing broadcast joins, one must be aware that if the broadcasted input is too large, "out of memory" errors can occur, due to the lack of memory to process all inputs. Hive 1.2.1, included in the Hortonworks Data Platform used for this book, automatically selects the most appropriate type of join. Nevertheless, since all configurations are set to their default values, Hive does not trigger a map join in the SF=300 workload (and for certain SF=100 queries as well), since the threshold regarding the fraction of memory dedicated for map join is probably surpassed. This causes a severe performance degradation for the star schema implemented in Hive. Such phenomenon raises a relevant discussion concerning the effect of the dimensions' size in the star schema modeling approach, which will be further discussed in subsection 8.2.3.

Besides these memory requirements, during the benchmark, we analyzed the cumulative and peak memories for each query running in Presto, and we observed that the star schema tends to achieve a higher peak memory when processing queries. The total amount of memory used for star schema-based queries is also substantially higher than flat-based queries in Presto's workloads. Regarding CPU usage, Figure 8-6 shows that in addition to being slower, the star schema tends to have a significantly higher CPU usage than a flat analytical object. On average, in Presto's workloads, the star schema uses considerably more CPU time. Consequently, higher CPU usage can also be seen as a drawback of star schema-based BDWs.

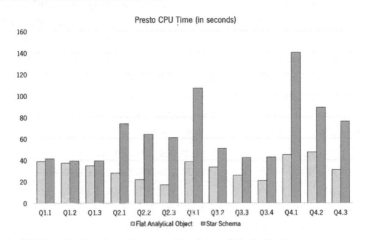

Figure 8-6. Presto CPU time for the star schema and the flat analytical object.

8.2.2. Improving Performance with Adequate Data Partitioning

Data partitioning can significantly impact the performance of storage systems. DWs are typically partitioned by date, or parts of a date (e.g., year, month, and day). However, there are other attributes that can be used for partitioning, related to specific implementation contexts. Depending on the attributes frequently used in the where clause of the queries, data partitioning can reduce query execution times, since the amount of data that needs to be processed will be much smaller. Another benefit of this technique is the simplification of CPE workloads, because one can make specific changes to previously loaded partitions without affecting the entire dataset. For example, if CPE workloads for sales data are executed each day, and there was a mistake in the data that was loaded yesterday, today's workloads can correct these mistakes by completely overwriting yesterday's partition without affecting the entire dataset. Sometimes, especially in Big Data environments, completely overwriting partitions is more efficient than updating multiple records. Furthermore, frequently, Big Data storage systems do not provide adequate updating capabilities (e.g., HDFS/Hive without ACID transactions enabled).

The workloads presented in the previous subsection do not rely on partitioning strategies, which is uncommon in real-world contexts. However, it allows to evaluate queries on large amounts of data. In certain organizations and contexts, even daily batches of data are significantly large and, therefore, it is relevant to understand how well flat analytical objects and star schemas can handle a large volume of data for certain infrastructures. In contrast, the workloads presented in this subsection use the SSB+ dataset partitioned by "order year", which is the attribute that appears more frequently in the where clause of the 13 SSB+ batch queries.

Figure 8-7 presents the results of the SF=300 workload using data partitioning, including a flat analytical object and a star schema, and comparing them with the results achieved in the SF=300 workload of the previous subsection. The performance advantage of using partitions is noticeable when the queries include the "order year" attribute as a filter. This is the main reason why the dataset was partitioned in the first place. This is true both for Presto and Hive. However, while Presto typically presents the expected behavior when the query does not benefit from the partition scheme, i.e., there is an increase in query execution or any difference is negligible, Hive presents an odd and unexpected behavior at first glance. The Q2 and Q3 variants are not supposed to benefit from this partitioning scheme, since they do not take advantage of any relevant "order year" filtering operations in the where clause. The Q3 variants tend to filter "order year" using a range of values, but the range is so wide that it is almost equivalent to scanning the entire dataset. Despite this, Hive's execution times for Q2 and Q3 variants (except Q2.2) drop drastically for the star schema using partitions, which is not expected at all.

After inspecting the execution of the queries more closely, we observe that using data partitioning, generally, some of the query plans for the Q2 and Q3 variants changed, and more mappers and reducers were produced. This number is affected by the organization of the ORC files in the system, as the partitioned Hive table may contain a different number of files with different sizes. Since the number of mappers and reducers is automatically derived in this work, this new number seems to drastically affect the query performance, resulting in a massive drop in query execution times for the star schema. These benefits are also present in the flat analytical object, but with much less predominance, since the query execution times for the Q2 and Q3 variants did not drop as significantly as in the star schema workload.

Presto does not behave like Hive for the star schema Q2 and Q3 variants, presenting expected results, i.e., results similar to the workload without partitions, even demonstrating small increases in execution times. This is to be expected, since there is the overhead of scanning multiple partitions when the query does not take advantage of the data partitioning scheme. In contrast, there is a significant performance advantage when running the Q2 and Q3 variants over the flat analytical object with partitions, which again is an unexpected behavior.

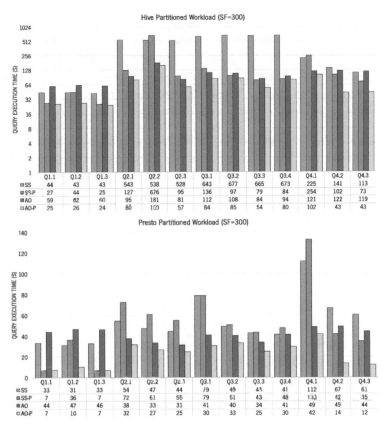

Figure 8-7. Large-scale batch SSB+ SF=300 workload with data partitioning.
Star schema (SS); star schema partitioned (SS-P); analytical object (AO); analytical object partitioned (AO-P).
Hive's results are based on E. Costa, Costa, & Santos (2017).

Investigating the issue in depth, we argue there are certain scenarios where natural hierarchies between attributes can cause ORC files to distribute data in such a way that it unintentionally improves query performance. This phenomenon happens because of a feature known as predicate pushdown at the ORC file/stripe level, together with file/stripe level statistics. For example, if one partitions a table by "supplier region", the queries that filter the data by "supplier nation" will also benefit from this partitioning scheme. The attributes "supplier region" and "supplier nation" form a natural hierarchy, and a specific partition will only contain countries that belong to the corresponding region. Consequently, the ORC files/stripes within this partition will provide statistics regarding the countries contained in them, and the query execution engine (e.g., Presto or Hive) can completely ignore files/stripes that do not contain the countries being filtered in the query, which makes query processing much faster, since it scans less data. However, this does not happen in the partitioning scheme used in this benchmark, as the Q2 variants do not filter the data by any attribute hierarchically related to "`order year`", and Q3.1, Q3.2, and Q3.3 only discard 1 in 8 years of data. Therefore, as previously explained, we can only conclude that the different organization of ORC files when using partitions may also affect stages, tasks, and drivers that are planned in Presto's queries, resulting in a performance boost, similar to the one caused by having different numbers of mappers and reducers in Hive, but with less predominance.

Overall, data partitioning is a mechanism that BDWing practitioners need to seriously take into consideration, as the performance advantage it brings is noticeable. One needs to understand recurrent query patterns, particularly the attributes that appear more frequently in "`where`" clauses, as well as understand specific needs for CPE workloads where data partitioning can be helpful.

8.2.3. The Impact of Dimensions' Size in Star Schemas

Large dimensions can have a considerable impact in star schema-based DWs, as they require more time to compute the join operations between the fact tables and the dimension tables. In previous workloads, one used larger dimensions' sizes. Although this may not be the usual scenario for many traditional contexts, such as store sales analysis, for example, larger dimensions are typically found in various Big Data contexts. Let us take into consideration a very large Web sales company such as Amazon, which has hundreds of millions of customers and products. In these contexts, the size of the dimensions may be very similar to those evaluated in subsection 8.2.1. In Big Data environments, there may be many other use cases that rely on very large dimensions, such as the set of Facebook users, which easily surpasses the 1 billion mark currently.

Nevertheless, there are many contexts where dimensions are small, because organizations can generate several sales transactions based on a small set of products, customers, and suppliers, for example. For this reason, it is interesting to analyze the performance impact caused by dimensions with different sizes. Figure 8-8 illustrates the results of the SF=300 workload for large and small dimensions.

The workloads for the flat analytical object were executed again, since smaller dimensions in the star schema also imply less cardinality in the descriptive attributes of a fully denormalized structure, e.g., if there are less rows in the customer dimension, there are also less distinct values in the "`customer name`" attribute. The cardinality of the attributes can also affect the performance of "`group by`" and "`order by`" operators and, therefore, the flat analytical object was reconstructed and evaluated again.

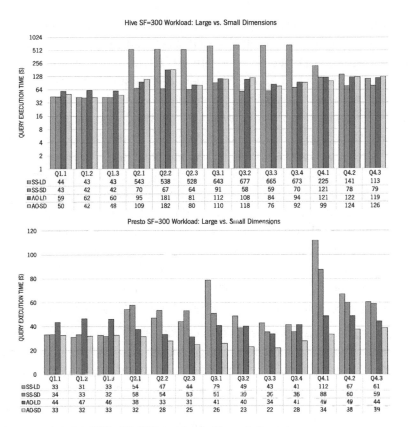

Figure 8-8. Large-scale batch SSB+ SF=300 workload with small dimensions.
Star schema with large dimensions (SS-LD); star schema with small dimensions (SS-SD); analytical object with large dimensions (AO-LD); analytical object with small dimensions (AO-SD). Adapted from C. Costa & Santos (2018).

At first glance, looking at Hive's workloads, the result was relatively unexpected. The flat analytical object, which until now was the modeling approach with the best performance in almost every query and workload, was surpassed by the star schema with small dimensions. This shows that when using Hive as the SQL-on-Hadoop engine, practitioners may sometimes benefit from modeling the BDW using dimensional structures, saving not only a considerably amount of storage size but, as Figure 8-8 demonstrates, also gaining considerable performance advantages. In this scenario, we conclude that if Hive is able to perform a map join, having a larger denormalized structure may not be appropriate for highly dimensional data, such as that generated by sales. The overhead caused by a storage size that is approximately 2.5 times larger (see Figure 8-4) causes a performance drop, and may turn into a bottleneck for the Hive query engine. Consequently, in these cases, practitioners may opt for the strategy presented in subsection 5.4.3, which discusses the modeling of traditional dimensions as complementary analytical objects for dimensional BDWing contexts.

© FCA

Considering Hive's results exclusively in a small dimensions scenario, they would favor the dimensional approach for modeling BDWs, particularly in the context of traditional DWing, where data is structured as fact tables and dimension tables (Kimball & Ross, 2013). Also, in the context of the proposed approach, where data is structured as analytical objects and complementary analytical objects. Consequently, given Hive's results, it sometimes makes sense to model parts of the BDW's data that way. However, Big Data often does not adequately fit into the strictures of dimensional and relational approaches (e.g., high volume/velocity sensor data or social media data). Taking a closer look at Presto's workloads, which are typically much faster than Hive's workloads, we observe that the star schema with smaller dimensions is significantly slower than the corresponding flat table. Furthermore, the star schema with smaller dimensions is frequently slower than the flat table with the higher attributes' cardinality (corresponding to larger dimensions). Overall, the star schema with smaller dimensions takes 61% more time to complete the workload when compared to the equivalent flat table. The discussion in this subsection illustrates why we use two SQL-on-Hadoop systems in each workload, as the insights retrieved from these specific tests sometimes differ depending on the system.

In summary, there are no strict rules to follow. In certain BDWing contexts, practitioners need to consider their limitations concerning storage size and the characteristics of a particular dataset. Relevant questions that need to be addressed are, for example: *is the data highly dimensional? Do the dimensions have a high number of rows or a large storage footprint and, therefore, not allow map/broadcast joins? Are these dimensions frequently reused by other analytical contexts?* In this book, these issues are discussed in subsection 5.4.3, which is the result of relevant insights provided by this evaluation. Furthermore, practitioners may need to perform some preliminary benchmarks with sample data before fully committing to either an extensive use of complementary analytical objects or flat analytical objects without any complementary joins.

8.2.4. The Impact of Nested Structures in Analytical Objects

Nested structures such as maps, arrays, and JSON objects can be significantly useful in certain contexts. Chapter 9 discusses one of these contexts, describing the implementation of a BDW for smart cities, wherein geospatial analytics is a priority, including several geometry attributes that are typically complex and nested. There are many contexts where nested structures can be seamlesly integrated in the data modeling approach. As exemplified in subsection 5.4.1, sales analysis is another context where practitioners may find the application of nested structures appealing, particulary using a less granular analytical object "orders" with the granularity key "order key", and using a nested structure to store the data about the products sold in a particular order (e.g., "product name", "quantity", and "revenue"). The proposed approach promotes the use of nested structures when feasible; however, it is important to ask oneself: *is this always the most efficient solution? Does processing less rows and having a smaller storage footprint always bring tangible advantages?*

To answer these questions, let us consider Figure 8-4, which compares the storage size (SF=300) of the different modeling approaches used in this work. Considering the nested analytical object created for this evaluation, we can conclude that it represents 68% of the equivalent flat analytical object's storage size, and roughly 186% of the equivalent star schema's size. Moreover, this new modeling approach reduces the number of rows from 1.8 billion to just 450 million, since the granularity of this analytical object is "order", instead of "lineorder". The data concerning the lines of the order is stored in a nested structure, namely an array of struct values called "lines" (similarly to the row datatype in certain SQL-on-Hadoop systems).

At first glance, these numbers look promising, but Figure 8-9 shows a different perspective with the results of executing the SF=300 Q4.1 in all modeling approaches. Q4.1 was chosen because it involves all the dimensions, representing a scenario wherein practitioners will need to aggregate and filter data that is stored in the nested attribute "lines". This allows to assess the effects of applying different operators to nested attributes, like the array of structs used in this context. Since one is dealing with nested structures, to achieve the same results as the remaining modeling approaches, other SQL operators, such as lambda expressions, must be used, as the following modification of the Q4.1 SQL query demonstrates:

```
SELECT od_year, c_nation, SUM(profit) AS profit
FROM (SELECT od_date, c_custkey,
        REDUCE(lines,
            CAST(0.0 AS real),
            (s, x) -> IF(x.s_region = 'AMERICA' AND (x.p_mfgr = 'MFGR#1'
        OR x.p_mfgr =
            'MFGR#2'), s + (x.revenue - x.supplycost), s), s -> s) AS profit
        FROM <db_name>.<table_name>)
WHERE c_region = 'AMERICA' GROUP BY od_year, c_nation ORDER BY od_year,
c_nation;
```

Despite saving storage space and having much less rows, the nested analytical object is the modeling approach with the lowest performance. It can be concluded that storing a large number of attributes in a complex structure may result in a large overhead for query processing times. After all, one is storing all the attributes of the "part" and "supplier" dimension in this array of structs, along with the various facts about the orders. Such data modeling choice requires lambda expressions or lateral views to answer Q4.1. In this particular test and looking into the query execution, Presto spends the majority of the time computing the lambda expression, which leads to a significant increase in query execution time.

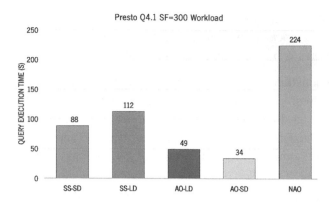

Figure 8-9. Performance of a nested analytical object in the SSB+ context.
Star schema with small dimensions (SS-SD); star schema with large dimensions (SS-LD); analytical object with large dimensions (AO-LD); analytical object with small dimensions (AO-SD); nested analytical object (NAO). Adapted from C. Costa & Santos (2018).

These results do not imply that processing nested structures is always detrimental for performance. It depends on the complexity of the structure and the kind of operators that will be applied. As shown in this evaluation, highly complex nested structures that are accessed sequentially to answer most queries may not be an adequate design pattern. However, as will be shown in Chapter 9, nested structures offer great flexibility, are significantly efficient for certain access patterns, and allow introducing new analytical workloads in the BDW, such as intensive geospatial simulations and visualizations.

8.2.5. Drill Across Queries and Window and Analytics Functions

In a traditional DWing context, submitting queries to combine data from multiple fact tables is a frequent phenomenon, which can also be described as drilling across fact tables. On the other hand, window and analytics functions (e.g., over clause, partition by, and rank) also play a relevant role in the *ad hoc* exploration of the data. Consequently, this subsection explores the performance of BDWs when using drill across and window and analytics functions, following the same strategies already presented above, i.e., using a flat analytical object and a star schema with Hive and Presto (with broadcast joins) as the SQL-on-Hadoop engines. Figure 8-10 summarizes the results for the four queries in this SF=300 workload, which are available in (C. Costa, 2017). The first three queries focus on drill across operations and the last one focuses on window and analytics functions, using the following questions:

- (Q5) Sum of the quantities ordered from the top 20 suppliers that had complaints from American customers within the last 4 years;

- (Q6.1) Number of times the company has sold parts from the manufacturer `'MFGR#3'`, provided by Asian suppliers, with an average selling price over 1000 USD in America, and which have been returned more than one time;

- (Q6.2) Top 10 parts with an average selling price over 1000 USD that were returned more than one time;

- (Q7) Top 5 parts of every market segment according to the generated revenue.

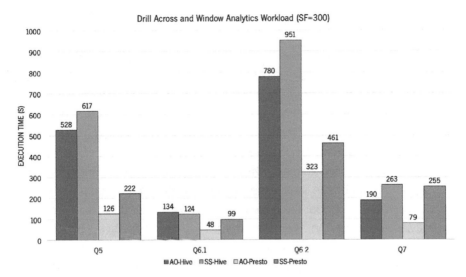

Figure 8-10. Performance of an analytical object (AO) and a star schema (SS) in a workload based on drill across queries and window and analytics functions.

The results of this workload revealed that, overall, the performance of a completely flat analytical object is more satisfactory than a star schema, although Q6.1-Hive is an exception. Considering Presto's results, which was the SQL-on-Hadoop engine that revealed greater differences, we can observe that the flat analytical object frequently completes the query in approximately half the time required by the star schema. Considering Hive's results, the star schema is faster in one of the four queries, although less significant than Presto's results (only 10 seconds). In contrast, considering Q6.2-Hive, one of the queries in which the star schema is slower than the flat analytical object, the difference in performance is noticiable. This shows that certain queries using subqueries with large (less filtered) intermediate results may significantly impact the performance when drilling across fact tables from a star schema. Consequently, we can conclude that, based on overall performance, flat analytical objects provide significantly lower execution times than star schemas in scenarios with drill across queries and window and analytics functions.

8.3. STREAMING OLAP

Streaming scenarios are common in BDWing contexts. The BDW must be able to adequately deal with the high velocity and frequency of the CPE workloads. Daily or hourly batch CPE workloads may not always be the most effective or efficient solution to solve specific problems, and streaming CPE workloads can be significantly useful in these cases. This section evaluates the performance of BDWs created using the proposed approach in streaming scenarios, while discussing several concerns that practitioners must take into consideration.

Using the SSB+ Benchmark, one can observe the performance of the streaming storage of the BDW. As illustrated in Chapter 5, there are several technologies that can be used to implement this storage component, which is responsible for storing data that flows continuously to the BDW with low latency requirements. In section 5.3, the trade-offs between these different technologies were also discussed. For example, Hive adequately deals with sequential access workloads, typically found in OLAP queries, but it is not adequate for random access, which is often suitable for storing streaming data. In contrast, NoSQL databases like Cassandra are efficient in random access scenarios, but typically fall short in sequential access workloads required for analytical contexts. Consequently, this subsection evaluates the performance of these two technologies using the two main data modeling strategies previously explored: a flat analytical object and a traditional star schema approach, as detailed in section 8.1. The data flow is the following:

1. A Kafka producer generates 10,000 records every 5 seconds.

2. A Spark Streaming application with a 10 seconds micro batch interval consumes the data for that interval and stores it in Hive and Cassandra.

3. Presto is used to query both streaming storage systems, every hour, over a period of 10 hours.

8.3.1. The Impact of Data Volume in the Streaming Storage Component

The performance of the streaming storage system of a BDW typically starts degrading as the amount of stored data increases. This is the main reason why the proposed approach includes an inter-storage pipeline to transfer the data from the streaming storage system into the batch storage system. Consequently, in this subsection, we are interested in analyzing how the data volume affects the performance of the streaming storage component of the BDW.

Figure 8-11 illustrates the total execution time for all streaming queries (Q8, Q9, and Q10) during a 10-hour workload with constantly increasing data volume. Each hour, all queries are executed using Presto, both for the flat analytical object and the star schema, and for Hive and Cassandra as well. The queries Q8, Q9, and Q10 focus on the following analytical questions:

- (Q8) The two countries that have the most positive average sentiment polarity united with the two countries that have the most negative average sentiment polarity;

- (Q9) The count of sentiments that were expressed by females in Portugal or Spain, grouped by product (`part`) category and period of the day (e.g., `dawn`, `morning`, `afternoon`, and `night`);

- (Q10) The groups of product (`part`) categories and genders with an average sentiment polarity greater than the total average sentiment polarity.

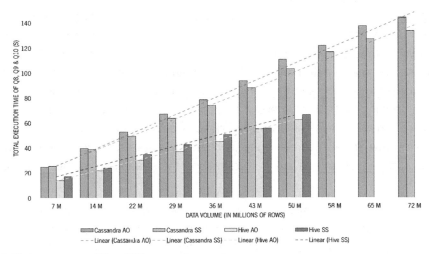

Figure 8-11. Cassandra and Hive SSB+ streaming results.
Star schema (SS); analytical object (AO). Adapted from C. Costa & Santos (2018).

There are several relevant insights that emerge from this evaluation. The first one concerns the overall effect of data volume on both systems. Looking into the trend exhibited by this 10-hour workload, we can conclude that both Hive and Cassandra are affected in a linear fashion, i.e., as hours pass by (and data volume grows), the increase in the execution time of the workload can be modeled as a linear function. A significant drop in performance as the data volume increases is to be expected in Cassandra, since as previously argued, sequential access over large amounts of data is not one of its strong features. However, this is not expected in Hive because, as the batch workloads demonstrate, when using Presto to query Hive, one is able to achieve much faster execution times than those obtained in this streaming workload, even with significantly higher SFs (e.g., SF=30 with 180,000,000 rows).

Despite this observation, which will be detailed afterwards, it can also be concluded that Hive is always considerably faster than Cassandra, until the mark of 50 million rows is reached. After this point, the Spark Streaming micro batch interval is too short for the demand, and the application also generates thousands of small files in HDFS (storage backend for Hive). Therefore, after the 50 million rows mark, hundreds of micro batches are being delayed, which makes the results for Hive inconclusive, as the number of stored rows does not match those in Cassandra. Overall, we can conclude that having small micro batch intervals when using Hive may severely deteriorate the performance of the system, complementing the conclusion made before regarding the overhead of having many small files stored in HDFS.

Cassandra also shows some delay in write operations when being queried by Presto, which causes the Spark Streaming application to queue a few micro batch jobs. However, this phenomenon is significantly less concerning than the one in Hive's scenario, and the streaming system is able to control the load without too much delay. Moreover, this is not caused by an increase in data volume, but rather a concurrency issue and resource starvation while Presto queries are running. One can always sacrifice data timeliness by increasing the micro batch interval, but comparing the results between Cassandra and Hive, the write latency and throughput should be identical. In this case, Cassandra adequately handles 20,000 rows every 10 seconds without significant delays, despite being slower, whereas Hive fails to do so, despite being faster for all workloads under the 58 million rows mark. This efficiency problem is discussed in more detail in the next subsection, among other relevant considerations for streaming scenarios in BDWing systems.

In this analysis, it is also interesting to evaluate the performance of a flat analytical object and a star schema. In this streaming context, the performance is relatively similar in both cases. The star schema is typically faster when using Cassandra, whereas the flat analytical object is typically faster when using Hive. In the SSB+ Benchmark, the star schema for the streaming scenario is not very extensive nor complex, which in this case favors this modeling approach, since queries do not have to join an extensive set of tables. Despite this, we can conclude that both modeling strategies are feasible, without any significant performance drawback. In the star schema's case, where the dimension tables are stored in Hive, we can also conclude that using a SQL-on-Hadoop system like Presto is also feasible for combining complementary analytical objects stored in Hive (e.g., "part" and "time") with streaming analytical objects stored in Cassandra. It is important to remember that the proposed approach uses the concept of complementary analytical objects to model dimensions, when practitioners prefer the use of dimensional structures for certain contexts (see subsection 5.4.3).

8.3.2. Considerations for Effective and Efficient Streaming OLAP

A successful streaming application can be seen as an adequate balance between data timeliness and resource capacity. To explain these trade-offs, this subsection is divided into three main problems that emerged from the evaluation of the SSB+ Benchmark, offering possible solutions to overcome these issues:

1. High concurrency in multi-tenant clusters (multiple users and multiple technologies) can cause severe resource starvation.

2. Storage systems oriented towards sequential access (e.g., Hive) may present some problems when using small micro batch intervals.

3. Inter-storage pipeline operations and CPE workloads should be properly planned, and the adequate amount of resources should be reserved.

Starting with the first problem, the proposed approach promotes a shared-nothing and scale-out infrastructure that is usually capable of multi-tenancy, i.e., adequately handling the storage and processing needs of multiple BDWing technologies and users. Streaming applications, like the one discussed in the previous section, normally require a constant allocation of CPU and memory for long periods of time. Data arrives at the system continuously, thus it needs to ensure that the workload has the required amount of resources available.

A common setup, like the one evaluated in this chapter, would comprise a producer (e.g., Kafka), a consumer (e.g., Spark Streaming), a storage system (e.g., Cassandra and Hive), and a query and OLAP engine (e.g., Presto). At first glance, the first three components of this setup may seem to work perfectly fine. However, once the query and OLAP engine is added, resource consumption can become significantly high, and the performance of the streaming application may suffer when the adequate trade-off between data timeliness and resource capacity is not chosen. Take as an example Figure 8-12. If carefully observed, in certain periods of time coinciding with the time interval when Presto queries are running, there is a significant increase in the processing time of the micro batches, thus causing an increase in the scheduling time of further micro batches.

In this case, this happens because there is not enough resource capacity in the current infrastructure to handle the processing demands of Spark Streaming, Cassandra, and Presto running simultaneously. In these periods of time, these technologies are competing for CPU usage, and the initial micro batch interval of 10 seconds is not enough to maintain the demands of the streaming application. Again, these insights bring us back to the previously discussed trade-off: either resource capacity is increased, in this case using more CPU cores, or the micro batch time interval is raised, which inevitably affects data timeliness. In this benchmark, the queries are only executed each hour, thus the system is only affected during these periods. However, in real-world applications, users are constantly submitting queries, which makes this consideration difficult to ignore.

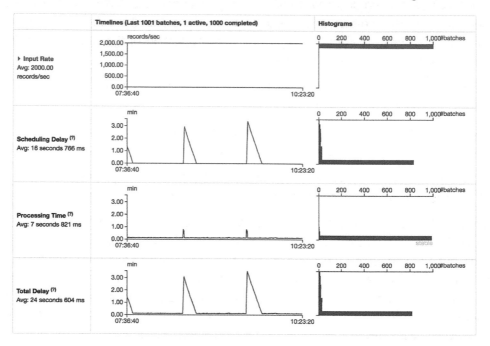

Figure 8-12. Spark Streaming monitoring GUI showing resource starvation when using Cassandra and Presto simultaneously. Adapted from C. Costa & Santos (2018).

As seen in the previous subsection, using storage systems like Hive for streaming scenarios has its advantages, specifically reduced query execution times, since this system can be considerably faster than Cassandra. Nevertheless, this performance advantage comes at a cost: as data volume increases, the number of small files stored in HDFS rises considerably, generating a significant load on the infrastructure. One small file is created for each RDD partition, in this case each 10 seconds, due to the chosen micro batch interval. In a matter of hours, the Hive table stores thousands of small files (see Figure 8-13). The problem is that, as the number of files increases, HDFS metadata operations take more time, thus affecting the time it takes for Spark Streaming to save the data in the Hive table. A write operation in HDFS includes steps like searching for existing files and checking user permissions (White, 2015); with thousands of files, this process can take longer than usual. Nevertheless, such problem can be solved by applying an adequate partition scheme to streaming Hive tables, e.g., partitioning by "date" and "hour", which creates a folder structure containing fewer files in each folder and, therefore, reducing the time to execute metadata operations (Vale Lima, Costa, & Santos, 2019).

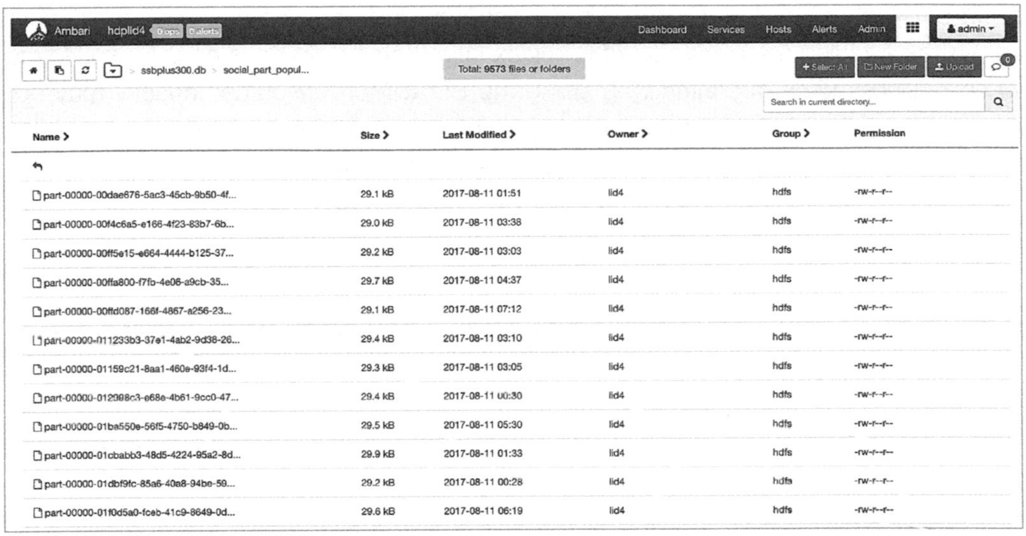

Figure 8-13. Thousands of small files created in HDFS (Hive's storage backend) when using Spark Streaming.

At this point, the system can be under intensive load and the Spark Streaming application queues hundreds of micro batches. Micro batches are queued when the Spark application cannot process them before the defined micro batch interval, in this case 10 seconds. Again, this predefined micro batch interval cannot ensure that the data is processed before the next micro batch, and the performance of the streaming application is thus compromised. In Hive's case, this is much more severe than the concurrency issue shown by running Cassandra and Presto simultaneously. In Hive's case, even increasing resource capacity is not the best solution, and one should opt for higher micro batch intervals, which will create bigger files. Moreover, the inter-storage pipeline is significantly relevant to periodically consolidate these small files into bigger files, or move them into another analytical object which contains historical data. It must be remembered that Hadoop prefers large files, which are then partitioned, distributed, and replicated as blocks.

Finally, and taking into consideration the phenomena discussed above, the inter-storage pipeline and CPE workloads should also be carefully planned when streaming applications are using the cluster's resources. These operations can be heavy on CPU and memory, and can unexpectedly cause resource starvation, as seen with Presto and Cassandra running simultaneously. Practitioners should not take this lightly. Linux Cgroups, YARN queues, and YARN CPU isolation can be extremely useful to ensure that the current infrastructure guarantees a rich, complex, and multi-tenant environment such as a BDW. These techniques make sure that resources are adequately shared by multiple applications, by assigning portions of the resources according to the expected workloads. Moreover, practitioners should evaluate their requirements regarding data timeliness, and avoid small micro batch intervals for streaming applications when they are not needed, as well as avoiding the execution of complex inter-storage pipelines or CPE workloads when business users are intensively using the BDW. More resource capacity may not always be the most efficient solution to the problem, since even in commodity hardware environments buying hardware always comes at a cost, whereas making some of these changes may increase efficiency without any significant cost.

8.4. SQL-ON-HADOOP SYSTEMS UNDER MULTI-USER ENVIRONMENTS

In real-world environments, the BDW will not be queried by a single business user. The system may have to support several decision-makers at different organizational levels. Single-user benchmarks allow us to understand the raw performance of a certain system without considering concurrency. However, one should design and implement a BDW capable of supporting several simultaneous users.

Since the proposed approach promotes the use of SQL-on-Hadoop engines as the frontend for querying and OLAP, the way these systems handle concurrency is a key factor for a BDW that adequately supports several users. Consequently, this section discusses the performance of Hive and Presto in a SF=30 concurrent workload, wherein four users execute the 13 SSB+ batch queries simultaneously. In this case, the smaller SF=30 was used to balance the concurrency requirements, the size of the dataset, and the available infrastructure (subsection 8.1.2). In this evaluation, the SQL-on-Hadoop systems were set to their default configurations.

Looking at Table 8.1, Presto stands out as the fastest engine. However, since one is looking into multi-user efficiency, execution time may not be the only metric to take into account. Obviously, if Presto is the fastest engine to retrieve the results to the concurrent users submitting the queries, it can perfectly be considered the most adequate system. The problem is that, in single-user workloads, Presto already tends to be significantly faster than Hive, which gives it a severe advantage in this multi-user test. Taking a closer look at Table 8.1, one of the most interesting insights is Hive's increase in execution time from single-user to multi-user queries. Despite its inferior performance in single-user workloads when compared to Presto, Hive is the system that gets less affected by having multiple users submitting queries simultaneously. An increase below the 3x mark means that, in a concurrent environment with four users, the system is able to execute the query faster than executing the same query four times in a single-user environment.

Table 8.1. Multi-user SSB+ workload SF=30 (in seconds).
Multi-user execution (M); single-user execution (S).

Queries	Hive (S)	Hive (M)	Presto (S)	Presto (M)
Q1.1	23	42	4	20
Q1.2	24	45	5	20
Q1.3	24	43	4	18
Q2.1	26	66	3	18
Q2.2	36	99	3	13
Q2.3	24	80	4	13
Q3.1	28	63	4	17
Q3.2	28	64	3	16
Q3.3	25	57	4	14
Q3.4	25	63	5	17
Q4.1	28	85	5	20
Q4.2	28	69	5	20
Q4.3	28	64	5	19
Total	347	839 (Increase: 1.4x)	52	225 (Increase: 3.3x)

These results aim to provide an overview regarding the performance of SQL-on-Hadoop systems under concurrent environments. Generally, it can be concluded that SQL-on-Hadoop systems are able to handle concurrent queries on relatively modest hardware, such as the one used for this book. Obviously, not all the configurations were changed, which does not always represent the best setup for these systems, especially since concurrency configurations are one of the aspects that may need some tuning to achieve optimal performance in production systems. However, performing a benchmark using the vanilla version of the systems also means that any kind of over-fitting does not occur and they are on the same level, without any misconfigurations. Depending on the SQL-on-Hadoop system practitioners end up choosing for their BDWs, it is advisable and necessary to read the documentation and adjust any relevant configuration. It must also be remembered that each version of the systems brings a number of improvements, and if concurrency performance is a critical factor for choosing the SQL-on-Hadoop system, then an on-site benchmark may be needed before making any decision.

9

BIG DATA WAREHOUSING IN SMART CITIES

INTRODUCTION

This chapter discusses the implementation of the SusCity BDW in the context of smart cities (C. Costa & Santos, 2017c; Monteiro, Costa, Pina, Santos, & Ferrão, 2018; SusCity, 2016), which is built upon the proposed models and methods as a demonstration case. In the context of smart cities, vast amounts of heterogeneous data are constantly being produced by an extensive network of interconnected things, including smartphones, smart meters, temperature sensors, noise sensors, smart appliances, location sensors, among many others. Moreover, there are also other data sources like the cities' transactional database systems, geospatial files, census data, and data provided by private companies responsible for certain city services. This phenomenon is typically associated with the concepts of IoT and Big Data (Jara, Bocchi, & Genoud, 2013). Consequently, smart cities are seen as rich BDWing contexts, given this extensive set of data sources and its relevance in the cities' decision-making process.

9.1. LOGICAL COMPONENTS, DATA FLOWS, AND TECHNOLOGICAL INFRASTRUCTURE

In the context of smart cities, an adequate BDWing approach is crucial to support the decision-making process at scale, complying with the characteristics of a BDW. Figure 9-1 presents the SusCity BDWing architecture, following the proposed approach. The logical layer helps researchers and practitioners understand the logical components of the system and how data flows throughout these components. It uses the taxonomy of the proposed approach, partially inherited from the NBDRA (NBD-PWG, 2015), since the lack of concepts standardization can be an issue in Big Data research. The technological infrastructure focuses on the technologies used for instantiating the logical components and on the infrastructure in which these technologies are deployed (detailed in subsection 9.1.2).

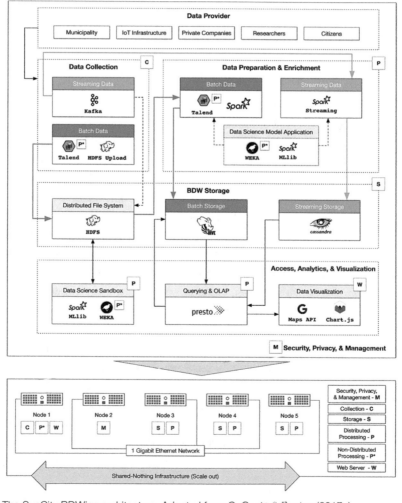

Figure 9-1. The SusCity BDWing architecture. Adapted from C. Costa & Santos (2017c).

9.1.1. SusCity Architecture

Regarding the logical layer, the first component is the data provider, which makes data available for further storage and processing. In a typical smart city context, which is the case of the SusCity research project (SusCity, 2016), the data provider component can include several actors:

- **Municipality** – the municipality itself can make available several data sources relevant for analytical tasks. For example, information about buildings or geospatial representations of the city's infrastructures. The city's transactional systems are also valuable data sources;

- **IoT infrastructure** – includes different kinds of sensors reporting electricity consumption, temperature, noise, and mobility patterns, for example. This data is relevant to understand events and real-time patterns in the city;

- **Private companies** – the city's infrastructures are not always public and, therefore, interactions with private companies are of major relevance in smart cities, for collecting historical energy consumption, buildings certificates, water consumption, census data, among many other data sources;

- **Researchers and citizens** – research projects conducted in the city are important data sources for the BDW; these include simulation data about different phenomena in the city (e.g., buildings' energy efficiency and mobility patterns), or any other relevant insights corresponding to scientific studies impacting the city. Moreover, citizens engaged in the initiatives promoted by researchers or by the municipality can provide useful data for the decision-making process, such as personal energy consumptions, mobility patterns, and service consumption habits.

Taking into consideration the data sources presented in Figure 9-1, data may arrive at the BDW via batch or streaming mechanisms. For data arriving in batches, we use Talend Open Studio for Big Data (Talend, 2017) and a HDFS client to upload it to the distributed file system. Before any preparation and enrichment process, data is first uploaded to HDFS, since raw data may serve further analytical purposes (e.g., training and testing data mining models) or may be useful for disaster recovery, should any problems occur in the storage component. HDFS is capable of handling large amounts of structured, semi-structured, and unstructured data, distributing them across several nodes in the cluster for further preparation and enrichment. For data arriving in a streaming fashion, Kafka (2018) is used to ensure highly scalable and robust data collection. Periodically, one can optionally move data from Kafka to HDFS, using systems such as LinkedIn's Gobblin (Qiao et al., 2015), in cases where streaming raw data is also useful for further purposes.

To prepare and enrich batch data, the SusCity BDWing architecture takes into consideration the volume of data. If the dataset being processed fits in the constraints

© FCA

of non-distributed technologies, Talend Open Studio for Big Data is used to prepare and enrich data, since it offers a wide set of processing components (e.g., filtering, aggregation, joins, and type parsing) in a user-friendly graphical interface. However, when the volume of the dataset requires distributed processing, Spark (Shanahan & Dai, 2015) is used to carry out preparation and enrichment tasks.

In streaming scenarios, one can use Spark Streaming to process data as it arrives at the BDWing system. Previously trained data mining models can also be applied in this phase, using WEKA (Hall et al., 2009) for small-scale algorithms (e.g., classification/regression of previously aggregated data or time series forecasting problems) and Spark MLlib for large-scale algorithms, i.e., when the training set contains vast amounts of very detailed data, not previously aggregated. Relevant data mining use cases in smart cities contexts may include forecasting and segmenting energy consumption (C. Costa & Santos, 2015); predicting attendance at the city's events; or segmenting buildings according to their characteristics and energy consumption. The same logic applies to unstructured data, since text mining algorithms, for example, can also be applied using WEKA, Spark or any other suitable technology, in order to extract structured patterns to be further stored in the BDW.

Once the data is prepared and enriched, it is stored in the storage component. The storage component comprises three subcomponents. The distributed file system (HDFS) acts as a staging area and sandbox, storing raw batch and streaming data, and temporary files needed in the data science sandbox component. It is a crucial component for ensuring a flexible storage capable of handling varied data from several sources. HDFS is also the underlying storage system for Hive, the technology used in the batch storage component of the proposed architecture. Hive is a DWing system on Hadoop, frequently mentioned as the *de facto* SQL-on-Hadoop solution. In the SusCity BDWing architecture, Hive tables stored as ORC files (Huai et al., 2014) are used to store large amounts of structured data, using the proposed data modeling method (the SusCity data model is presented in section 9.2). For this book, Hive only stores data arriving in batches, since it is mainly designed for fast sequential access to data. For fast random access, Cassandra is used as the NoSQL database supporting the streaming storage component, since one can ensure hundreds or thousands of concurrent writes frequently required by typical streaming applications. Previous benchmarks reveal that Cassandra is a suitable distributed database for intensive random read and random write scenarios (C. Costa & Santos, 2016b). Moreover, Cassandra tables are also modeled according to the data modeling method proposed in this book (section 9.2).

Regarding the use of Hive as a streaming storage system, if practitioners overlook some configuration and management aspects (e.g., ensuring efficient compaction techniques), Hive tends to generate several small files in HDFS, which can become a bottleneck in the system. Despite the current advancements regarding Hive's transactions (Apache Hive,

2018), Hive's suitability for a large number of concurrent and continuous writes needs to be tested in prototype or production systems. As mentioned, Hive is a DWing system on Hadoop mainly used to scale OLAP applications. Since NoSQL databases are mainly designed to scale OLTP applications (Cattell, 2011), they are not as effective and efficient as Hive in sequential access scenarios, typically required by OLAP applications, as can be further seen in section 9.3. Therefore, the SusCity BDWing system considers these trade-offs and maintains two separate storage technologies for batch and streaming data. As techniques and technologies evolve and stabilize, the same technology may be able to adequately support both scenarios, as discussed in section 5.3.

The goal of the BDW is to support analytical tasks. Consequently, the access, analytics, and visualization component is crucial to deliver adequate insights for data-driven decision-making processes. The querying and OLAP component, using Presto, ensures an adequate communication between the batch storage, the streaming storage, and the data visualization component. Presto was open sourced by Facebook, and it is considered as a SQL-on-Hadoop system providing low latency query execution over large amounts of data (Presto, 2018). In fact, it is more than a SQL-on-Hadoop system, since it can provide a SQL interface for a vast set of storage technologies besides Hive (Hadoop), including NoSQL databases like Cassandra and MongoDB. Therefore, in the proposed architecture, Presto is used to query Hive tables and Cassandra tables. As previously discussed, since Cassandra is less efficient than Hive for fast sequential access (section 9.3), we also use Presto to transfer data between Cassandra and Hive, avoiding the accumulation of vast amounts of historical data in the Cassandra tables. For more complex queries that surpass the interactivity threshold defined for the SusCity data visualization component (10 seconds), Presto is also used to create materialized views stored in Hive tables.

Although several improvements have been made in Hive, such as the Tez execution engine (Floratou et al., 2014; Huai et al., 2014), Presto achieves significantly faster execution times when querying Hive tables (as can be seen in Chapter 8), hence why it is used as the querying and OLAP engine in the proposed architecture. Interactive query execution is one of the main requirements of the SusCity data visualization component, in order to engage users through a responsive interface. Therefore, the data visualization component (discussed in section 9.4) uses Presto to submit SQL queries to the batch and streaming storage systems. Presto can also combine data from these two components using a single query (e.g., joins and unions), which is of major importance for combining historical and streaming data into a unified view of the data.

Although some benchmarks demonstrated that interactive SQL-on-Hadoop systems similar to Presto (e.g., Impala) may struggle with datasets that do not fit into memory (Floratou et al., 2014), we did not feel the need to use the Hive execution engine in the SusCity project. Presto was able to execute all the workloads requested by the SusCity testbed, processing several Gigabytes of data from energy grid simulations, buildings information, geospatial files, historical energy consumption data, and more than one hundred smart meters. However, if certain scalability issues arise, the Hive execution engine is always available for more demanding workloads. Scalability will certainly not be an issue, since Presto is being used by Facebook to perform queries over its Petabyte-scale Hive DW, thus it is possible to scale the cluster to accommodate growing data in a smart cities context.

The application of data science models (e.g., data mining and text mining models) is only possible with an adequate sandbox where data scientists can explore the data, training and testing models to support their hypotheses (C. Costa & Santos, 2017b). Therefore, the proposed architecture includes a dedicated component, named data science sandbox, which interacts with HDFS. Since raw batch and streaming data can be stored in HDFS, data scientists can interact with this data to produce models capable of extracting patterns and making predictions when new data arrives at the preparation and enrichment component. WEKA and Spark are the driving forces for this purpose, as previously discussed.

Security, privacy, and management constitute a relevant component in the SusCity BDWing architecture. There are certain Hadoop-related technologies that can be used to ensure a secure environment that is properly managed. For this book, Kerberos is used to provide a secure authentication protocol in Hadoop. For an extra-layer of security and privacy, Ranger can be used to deploy rigorous authorization policies, defining which users have access to certain files or tables (Hortonworks, 2016). Regarding Cassandra's security, TLS/SSL encryption can be used for client-to-node or node-to-node communications. Cassandra also provides simple password authentication and an internal authorization model. In a smart city context, data privacy is a main concern and, therefore, whenever possible, we encourage anonymization of sensitive data before storing it in the BDW (C. Costa & Santos, 2016a). Finally, Ambari can be used to manage and monitor the Hadoop components deployed in the cluster (Apache Hadoop, 2018).

9.1.2. SusCity Infrastructure

All the components and technologies discussed in the previous subsection are deployed in 5 commodity hardware machines installed on-premises, which have been capable of supporting the workloads demanded by the SusCity testbed. In this demonstration, each machine has 16GB of RAM, 1 Intel Core i5 quad-core CPU, and 500GB 7,200rpm hard disks (except node 1, which has an Intel Core i7 quad-core CPU and a 256GB SSD). All the machines are connected using a 1 Gigabit Ethernet switch and Ethernet CAT6 cables, since Hadoop clusters should be deployed using at least a 1 Gigabit Ethernet network (Shvachko et al., 2010). The use of Big Data technologies like Hadoop, Spark, and

Cassandra, which rely on commodity hardware and shared-nothing infrastructures, allows to scale the cluster as data volume increases and workloads become more demanding.

Due to resource limitations, Hadoop/Hive and Cassandra nodes are co-located (node 3, node 4, and node 5). Presto and Spark are also deployed in these nodes for co-located processing. Distributed processing technologies should be co-located with storage nodes, since in Big Data environments the processing should be brought closer to the storage, to avoid moving large amounts of data through the network (C. L. P. Chen & Zhang, 2014). The collection technologies (Kafka, Talend Open Studio, and HDFS client), non-distributed processing technologies (Talend Open Studio and WEKA), and the Web Server providing dashboards with Chart.js (2017) and the Google Maps API (Google Maps, 2017) are all deployed within node 1, again due to resource limitations. Ideally, in a production environment, these elements should have dedicated nodes and Kafka should be distributed across several nodes in the cluster. Node 2, besides being the Hadoop NameNode, enables tasks related to the security, privacy, and management of the cluster, containing the Kerberos Key Distribution Center and Ambari.

9.2. SUSCITY DATA MODEL

The main concept in the proposed data modeling method is the analytical object, representing a subject of interest to be analyzed. Typical analytical objects in smart cities may include: general indicators about buildings; buildings energy consumption; losses in the energy grid; indicators about the nodes in the energy grid; and the energy consumption recorded by smart meters.

Making an analogy with traditional DWs, analytical objects have the same capabilities as fact tables. But unlike fact tables, they typically are fully denormalized structures, in which all the attributes needed for analyzing the subject of interest are included in one single analytical object, without the need for dimension tables, avoiding constant and demanding join operations. Join operations in Big Data environments are costly (Floratou et al., 2014; Marz & Warren, 2015; NBD-PWG, 2015; Wang, Qin, Zhang, Wang, & Wang, 2011), since tables may store vast amounts of data. Joining several dimensions with fact tables for each query can be significantly resource-demanding (see subsections 5.4.2 and 5.4.3 for concepts focusing on efficient dimensional patterns and join operations, and subsection 9.2.1 for their specific application in the SusCity BDW).

Guided by the approach proposed in this work, each analytical object of the SusCity data model contains two types of attributes: descriptive attributes (top half of the analytical objects in Figure 9-2) and analytical attributes (bottom half of the analytical objects in Figure 9-2). Moreover, outsourced descriptive families (see subsection 5.4.3) for each analytical object are also presented in Figure 9-2. Descriptive attributes support

© FCA

typical OLAP tasks by providing different perspectives for aggregations and filtering operations. These are analogous to the attributes of the dimension tables in traditional DWs, and allow to interpret the analytical attributes through different perspectives. Analytical attributes are analogous to the facts in a traditional fact table, but with the particularity that they can not only contain facts (historical indicators), but also predictions derived from applying of data science models. Take as an example the analytical object "buildings energy consumption" in Figure 9-2, which contains a cluster defining the consumption behavior of each building and a forecast of its energy consumption (kWh) for the following days, information obtained using the WEKA's clustering and time series forecasting algorithms, as proposed in (C. Costa & Santos, 2015).

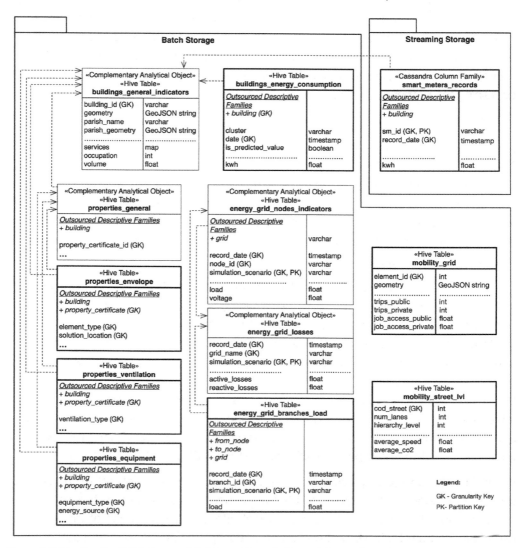

Figure 9-2. The SusCity BDW data model. Adapted from C. Costa & Santos (2017c) with extended content.

Descriptive and analytical attributes can store simple data types (e.g., integer, float, and varchar) or complex types (e.g., arrays, maps, and GeoJSON objects). The use of complex types to extend the capabilities of BDWs is detailed in subsection 9.2.2. Descriptive attributes are also relevant to define partition keys and granularity keys. Some descriptive attributes can be used to control certain aspects of data locality. In Hive, a partition key distributes the data throughout different folders according to the value of the attribute. In Cassandra, the partition key helps determining which node should be used to store/read the data, since it tries to evenly spread the data across different nodes in the cluster. There is no rigorous rule for defining partition keys, and one should evaluate the patterns of the queries and/or the refreshing rates of the analytical objects. For example, in the SusCity BDW, we use the "simulation scenario" attribute for the partition key in the "energy grid losses" and "energy grid nodes indicators" Hive tables, since one loads the data in batches corresponding to yearly simulations for each stress scenario in the energy grid. Moreover, the "simulation scenario" is an attribute frequently used in the where clause of the queries that refer these analytical objects. Regarding the "smart meters records" Cassandra table, it is possible to use "sm id" as the partition key, balancing the data throughout the nodes in the cluster according to the identifier of the smart meter. In Cassandra, the partition key is the first part of the PK (read primary key), which in this case is a compound key using "sm id" and "record date". In Hive, one does not need to define PKs (read primary keys).

As can be seen in Figure 9-2, analytical objects can be joined together to answer certain queries. These join operations are optional, because analytical objects can be modeled without any external references to other objects, which is the case for most of the analytical objects in the SusCity BDW. There is no need for declaring FKs at object creation, and analytical objects can be joined using all the attributes whose values match. This approach is detailed in subsection 9.2.1.

Complementing what was already mentioned regarding querying and OLAP, it is possible to use Presto to perform joins and unions between batch analytical objects (Hive tables) and streaming analytical objects (Cassandra tables), providing useful insights extracted from historical and real-time data, as can be seen in Figure 9-2. The size of the tables being joined is of major relevance for an adequate query performance, and it should be taken into consideration. This is the reason analytical objects should not be joined in their raw state. First, one needs to aggregate and filter (as much as possible) each analytical object involved in the join operation. The larger the inputs on each side of the join operation, the more complex and slower the query becomes. In this case, materialized views stored in Hive tables are significantly helpful for maintaining interactive response times in query execution and the responsiveness of the data visualization platform (see subsection 5.4.2 for more details on join operations and materialization processes).

Regarding this modeling approach, we can highlight three major strategies: using fully denormalized structures to avoid the cost of join operations in Big Data environments; using nested structures, which are not typically found in traditional modeling techniques, can provide more flexibility and performance advantages in specific scenarios; dividing data flows and storage components into batch and streaming, as discussed by Marz and Warren (2015), but that does not imply following different data modeling strategies, as the proposed data modeling method demonstrates, since it can be used both for batch and streaming contexts.

9.2.1. Buildings Characteristics as an Outsourced Descriptive Family

Looking at Figure 9-2, it can be seen that the Hive table "buildings general indicators" can be joined with the "buildings energy consumption" table. This capability is useful to understand relationships between the characteristics of buildings and their energy consumption, addressing questions such as: *to what degree does the number of occupants influence the building's energy consumption?*

These join operations are different from the join operations required between fact tables and dimension tables, since one only uses them in queries that relate different analytical objects, which is less frequent than joining fact tables and dimensions for each query. Considering the SusCity BDW, the "building id" attribute is present, among others, in three analytical objects: "buildings general indicators"; "buildings energy consumption"; and "smart meters records". Instead of replicating the information about buildings in these three objects and creating unnecessary redundancy, taking into consideration that the "buildings general indicators" object is relatively small (around 60,000 records for the city of Lisbon), it can be easily joined with the other two objects, to answer specific questions. Consequently, one can outsource the buildings characteristics to the "buildings general indicators" complementary analytical object, and only place the "building id" in the "buildings energy consumption" and "smart meters records" analytical objects as a link to the outsourced descriptive family (similarly to a FK in traditional DWs).

Nevertheless, when there is no need to relate energy consumption with buildings characteristics, all three objects are completely independent and can answer different queries without relying on any join operation. Such flexibility is one of the strongest points of the proposed approach, which provides constructs and structured guidelines that practitioners can follow to solve specific problems, depending on the considerations and trade-offs previously discussed in section 5.4.

9.2.2. Nested Structures in Analytical Objects

In the SusCity BDW, complex types are used to store nested structures that will be later interpreted by the data visualization component. For example, a large building can be associated to more than one service (e.g., laundry, supermarket, restaurant, and gym). Using a nested complex type like a map (e.g., HashMap), one can store this data using a single record associated with that building. Saving geometry objects in GeoJSON strings is also relevant for geospatial analysis in a smart city context. Figure 9-3 exemplifies this data modeling technique, showing how the "services" map and the "geometry" GeoJSON are stored. As can be seen in Figure 9-3, following the proposed approach, one can arrange the number of services by distance and type nested in the "buildings general indicators" analytical object, without having to create a new object to store the services for each building. This provides significant flexibility when creating custom-made data visualizations (see subsection 9.4), avoiding the need to perform complex queries to join different tables.

building_id	geometry	services
PN1002_Bld1006	{"coordinates":[[[[-9.095774715532317,38.75531748705829,0.0], [-9.09581794753689,38.754789659895685,0.0],[-9.09645431644,38.75508346464763,0.0], [-9.096450674839106,38.75512791869151,0.0],[-9.095863311171968,38.754856573974815,0.0], [-9.09582372031163,38.755340111999075,0.0], [-9.095774715532317,38.75531748705829,0.0]]],"type":"MultiPolygon"}	{"Closest":{"bus_station":1," restaurant":1," transit_station":1}}
PN1017_Bld1014	{"coordinates":[[[[-9.097215529256525,38.75555230452945,0.0], [-9.097043511328877,38.75554367148119,0.0],[-9.097049142417907,38.7554749032351,0.0], [-9.097229969051648,38.75548398426785,0.0],[-9.09725216370908,38.75548509831462,0.0], [-9.097274238654201,38.75548620703925,0.0],[-9.09726356044447,38.75547028727257,0.0], [-9.09734857452456,38.755359804547815,0.0],[-9.097422656424303,38.75539441765938,0.0], [-9.097316956110943,38.75553312604815,0.0],[-9.097299143785282,38.75555650063938,0.0], [-9.097215529256525,38.75555230452945,0.0]]],"typo":"MultiPolygon"}	{"Intersect":{"pharmacy":1," electronics_store":1," store":2}}
PN1005_Bld1020	{"coordinates":[[[[-9.097635848103682,38.75565135861262,0.0], [-9.09772348462946,38.75553497185145,0.0],[-9.098142617521873,38.75572847040772,0.0], [-9.09805498146341,38.75584485747365,0.0], [-9.097635848103682,38.75565135861262,0.0]]],"type":"MultiPolygon"}	{"Intersect":{"school":1}}
PN1005_Bld1024	{"coordinates":[[[[-9.098588536159713,38.75593433195039,0.0], [-9.098610737448173,38.75590484691583,0.0],[-9.098832011608275,38.756006999134215,0.0], [-9.098809810380983,38.75603648420953,0.0], [-9.098588536159713,38.75593433195039,0.0]]],"type":"MultiPolygon"}	{"Closest":{"school":1}}

Figure 9-3. SusCity nested structures (example).

9.3. THE INTER-STORAGE PIPELINE

The need to transfer the data between storage components was already highlighted in section 9.1, but it will be quantitatively evaluated in this section. As mentioned, NoSQL databases are OLTP-oriented (Cattell, 2011), unlike Hive, which is an OLAP-oriented technology. Typically, OLTP systems relax sequential access efficiency for random access efficiency. Therefore, systems like Cassandra are suited for constant random write operations frequently demanded by real-time collection of data from thousands of smart meters. However, these systems lack efficiency to process (e.g., aggregate) large amounts of historical data, which is frequently demanded by OLAP queries. Table 9.1 presents the results from an experiment conducted in the infrastructure and testbed of the SusCity research project, evaluating the response times when submitting Presto queries to Hive tables and Cassandra tables. As demonstrated in Table 9.1, given the same analytical object and the same amount of data, Presto OLAP queries on Hive tables (ORC file format) perform significantly faster than the queries on Cassandra tables. This corroborates previous observations and the decision to periodically move historical data from Cassandra to Hive, maintaining only the most recent data in Cassandra. The periodicity of this data transfer depends on the specific requirements regarding interactivity in response times, the volume of data being stored, and the available infrastructure. In this setup, data can be transferred on an hourly, daily, weekly or monthly basis, for example.

Table 9.1. Performance comparison between analytical objects stored in Hive and Cassandra. Based on C. Costa & Santos (2017c).

Query	Input Rows	Output Rows	Hive	Cassandra
Show the last 10 smart meters records	~2.8 million	10	0.56s	3.08s
Calculate the average of kWh grouped by smart meter	~2.8 million	214	0.56s	4.2s
Count how many records a certain smart meter contains	~2.8 million	1	0.74s	0.98s

9.4. THE SUSCITY DATA VISUALIZATION PLATFORM

Throughout this chapter, we focused on the logical and physical layers of the BDW. In this section, we highlight some relevant use cases in which the data visualization platform can help the city's stakeholders in the decision-making process. As previously noted, the SusCity data visualization platform was developed using modern JavaScript libraries like the Google Maps API V3 and Chart.js. Obviously, since it is a Web-based platform, core languages are also present (HTML, CSS, and pure JavaScript), as well as other supporting JavaScript libraries like jQuery (2017). It is a platform purely based on a service-oriented architecture, using Java REST Web services to ensure the communication between the JavaScript components of the platform and the querying and OLAP engine instantiated by Presto. Each query submitted to the BDW goes to this REST backend for an adequate modularity of the platform. Using this service-oriented and modular approach, it is easier to update or replace components and technologies, if that need arises in the future.

In this section, we will present various dashboards developed in the SusCity research project, which also represent interesting applications for other smart cities initiatives. The following dashboards are just a few examples of the SusCity data visualization platform's capabilities, and the SusCity demonstration case itself considers other data sources and experiments (e.g., data mining and machine learning insights) that were not fully developed and implemented in the visualization platform. Furthermore, due to security and privacy issues, the visualizations illustrated in this section are built upon incomplete, omitted and/or changed testbed data and, therefore, results are not conclusive for any real-world based decision-making process.

9.4.1. City's Energy Consumption

The first dashboard (Figure 9-4) is based on the energy consumption of each parish in the city (two parishes in the SusCity testbed). Decision-makers can understand the energy consumption in each parish and analyze the city's consumption by the hour, time period (e.g., morning or afternoon) or by quarter. Users can interact with multiple parishes by clicking on them, revealing the energy consumption for specific parishes, and comparing it with the overall consumption of the city, with the goal of extracting insights regarding critical zones in the city. In the SusCity data visualization platform, one can also make available the predictive capabilities of the SusCity BDW, such as the segmentation (clustering) of buildings according to their energy consumption, and the respective energy forecasting for the next days or weeks, as conceptually explored in this book (section 9.2) and as also discussed in C. Costa and Santos (2015).

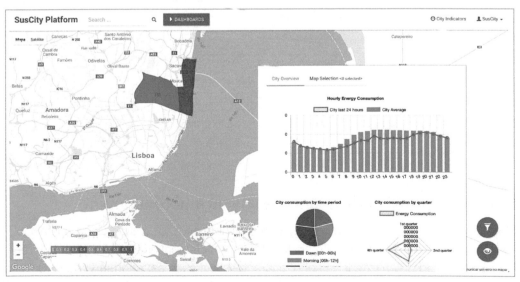

Figure 9-4. SusCity data visualization platform – energy consumption dashboard. Adapted from C. Costa & Santos (2017c).

9.4.2. City's Energy Grid Simulations

A dashboard to simulate and analyze stress scenarios in the energy grid can be significantly useful in the context of smart cities, as depicted in Figure 9-5. Each scenario corresponds to a set of input parameters (e.g., number of electrical vehicles, photovoltaic area, and number of charging stations) that may affect the behavior of the energy grid, such as energy losses, load, and maximum peak power. In the SusCity BDW, the results of the simulations for these scenarios are stored in analytical objects, as presented in section 9.2, and the data visualization platform can use the querying and OLAP engine to extract and provide useful insights for stakeholders interested in the impact that certain initiatives have on the energy grid. Due to the modular and service-oriented nature of the SusCity data visualization platform, and the flexible and scalable SusCity BDW, one can provide dashboards for decision-makers, regardless of data volume, variety, and velocity, without being held back by rigid data modeling techniques and complex data CPE pipelines.

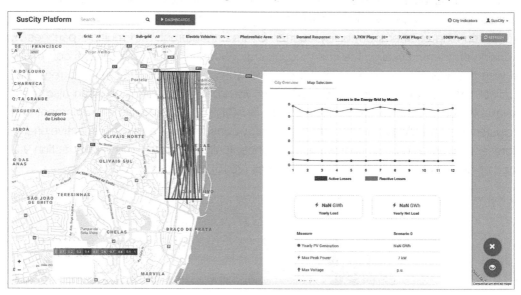

Figure 9-5. SusCity data visualization platform – energy grid simulation dashboard. Adapted from C. Costa & Santos (2017c).

9.4.3. Buildings' Performance Analysis and Simulation

Understanding the buildings' efficiency is a crucial aspect for a smart and sustainable city. One of the SusCity platform main focuses is the geospatial analysis of the buildings in Lisbon, based on an extensive set of characteristics, such as: geometry; construction; energy consumption and efficiency; envelope properties (e.g., type of window and type of window frame); heating and cooling systems in use; and occupation schedule.

The flexible data model and storage components of the SusCity BDW, i.e., the lack of a strict relational data model and the efficient use of GeoJSON objects, together with a rich API for geospatial analytics like the Google Maps API, provide an extensive set of analytical capabilities for the city's government. As can be seen in Figure 9-6, stakeholders can visualize the general distribution of the energy classes across Lisbon, the thermal inertia of the buildings, type of window and window frame, among other metrics georeferenced by building. Each chart is interactive and can be used as a filter to analyze how buildings are related to a certain property (e.g., metal window frame, double glass window, and low thermal inertia). As a consequence, the dashboard in Figure 9-6 is not only useful for the city's government, also for private companies interested in promoting retrofitting initiatives to modernize buildings.

Figure 9-6. SusCity data visualization platform – buildings analysis dashboard.

Similarly to the energy grid simulations (subsection 9.4.2), the SusCity BDW can also support simulations at the building level. Figure 9-7 demonstrates the use of simulation data to evaluate the impact of specific retrofitting initiatives (e.g., change the windows in 25% of the buildings in Lisbon). Besides analyzing the impact of retrofitting initiatives, decision-makers can also use the dashboard in Figure 9-7 to analyze the archetype of a specific building, among many other characteristics previously mentioned: geometry; construction; heating and cooling systems; and occupation schedule. Consequently, having a BDW whose data model facilitates the integration of a vast set of data sources, without rigid structures, is one of the main aspects that allows the development of these Big Data analyzes in the context of smart cities.

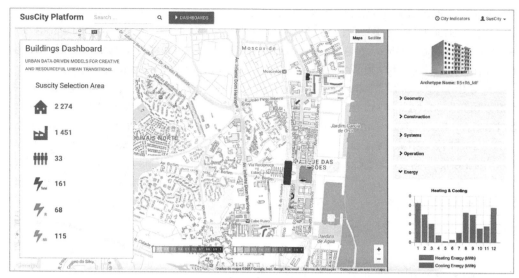

Figure 9-7. SusCity data visualization platform – buildings simulation dashboard. Adapted from Monteiro et al. (2018).

9.4.4. Mobility Patterns Analysis

Studying mobility is a crucial aspect in a smart city. Understanding how people or goods travel within the city, or how citizens tend to use private or public transports, for example, is an interesting subject for decision-makers, including the city's government and public/private transportation companies. Another interesting scenario is the footprint analysis of the city's streets, according to several indicators, such as CO_2 emissions or average speed.

Figure 9-8 focuses on the analysis of the city's mobility patterns, to foresee future initiatives to facilitate the use of either private or public transportation. The analysis follows these steps:

1. The city is modeled as a grid with several sections.

2. Every section is colored in the map according to one of three indicators: number of daily trips; job transport accessibility; and average travel time.

3. By clicking in one section of the grid, decision-makers can understand how that section performs regarding the three indicators mentioned above.

Analyzing the data at the grid level is interesting for understanding the mobility behavior within different sections of the city. However, the flexible data model of the SusCity BDW and the geospatial capabilities of the SusCity data visualization platform also allow the analysis of several indicators at the street level. To demonstrate these capabilities, Figure

9-9 presents several streets colored according to the ratio between average speed and maximum allowed speed, so that stakeholders can understand in which streets citizens tend to drive over the speed limit. The analysis in Figure 9-9 focuses on security concerning mobility patterns, but there are several other potential use cases for this kind of analysis, such as identifying the busiest or more polluting streets, for example.

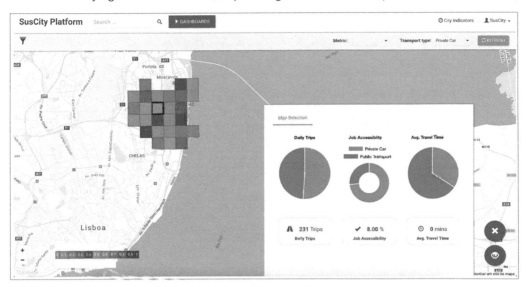

Figure 9-8. SusCity data visualization platform – mobility grid dashboard.

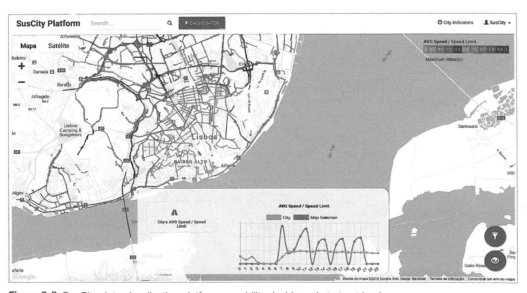

Figure 9-9. SusCity data visualization platform – mobility dashboard at street level.

10 CONCLUSION

Throughout this work, Big Data proved to be a concept of major relevance in today's world, whose popularity has increased considerably during the last years. Areas like smart cities, manufacturing, retail, finance, software development, environment, digital media, among others, can benefit from the collection, storage, processing, and analysis of Big Data, leveraging unprecedented data-driven workflows, and considerably improving decision-making processes. This new type of data is being defined not only by its characteristics (e.g., volume, variety, and velocity), but also by the limitations it imposes on traditional storage and processing technologies. Organizations seeking Big Data initiatives face several challenges, such as the lack of consensus in definitions, models, and architectures, and difficulties concerning the Big Data life cycle design and implementation.

Since the DW concept has a long history as one of the most valuable enterprise data assets, this book aimed to describe its role, design, and implementation in Big Data environments. The concept of BDW is emerging as either an augmentation or a replacement of the traditional DW. Research in this topic is still in its infancy, and as Big Data is often synonymous with ambiguity, the same happens for the concept of BDW. After analyzing the scientific and technical literature, this book shows that the BDW can be defined using the following characteristics:

- Parallel/distributed storage and processing of large amounts of data, including fault-tolerance issues;

- Scalability (accommodate more data, users, and analyzes) and elasticity, using commodity hardware to lower the costs of implementation and maintenance;

- Flexible storage, including unstructured data;

- Real-time capabilities (stream processing, low latency, and high frequency updates);

- High performance with near real-time response;

- Interoperability in a federation of multiple technologies;

- Mixed and complex analytics (e.g., *ad hoc* or exploratory analysis, data mining, text mining, statistics, machine learning, reporting, visualization, geospatial analytics, advanced simulations, and materialized views).

Considering the state of the art in BDWing, we conclude that there is no common approach to building BDWs. There are numerous Big Data technologies to choose from, each trying to stand out, which creates barriers in the design and implementation of Big Data solutions like BDWing systems, since these technologies' role is mostly misunderstood, eventually overlapping each other. Current logical architectures and non-structured contributions only solve part of the problem by providing some general and relatively unstructured constructs and guidelines. However, ambiguity regarding the BDW techniques and technologies that are more adequate for several contexts still prevails, mainly due to the lack of general-purpose, detailed, integrated, and adequately evaluated approaches.

Currently, the design and implementation of BDWs is mainly seen as a use case driven approach, instead of a data-driven one, which used to be the case for traditional DWs. Previously, data modeling was the primary concern, but, nowadays, practitioners are mainly concerned with finding the right technology to meet the demands of Big Data, leading to possible uncoordinated data silos. It would be a mistake to discard years of architectural best practices based on the assumption that storage for Big Data is not driven by data modeling (Clegg, 2015). Works related to the SQL-on-Hadoop movement are a suitable proof that the data structures known for a long time are still relevant, although modified and optimized. Obviously, unstructured data does not adequately fit into these structures, but, as this book has demonstrated, there are data science techniques for extracting value from data and subsequently fuel the BDW (e.g., data mining and text mining). Complex systems like BDWs require changes in different logical and technological components, data flows, and data structures, but this does not imply discarding previously useful models and methods in favor of a use case driven approach.

Until now, there was no structured and general-purpose approach describing how to design and implement BDWs, with adequately evaluated models (representations of logical and technological components, data flows, and data structures), methods (structured practices), and instantiations (e.g., demonstration cases through prototyping and benchmarking). This scientific and technical gap was the main motivation for this book. In our opinion, the existing logical architectures, non-structured guidelines, best practices, and implementations in specific contexts, although relevant, do not provide the complete, general-purpose, detailed, and thoroughly evaluated approach that practitioners require to design and implement BDWs according to their characteristics. The gap between *this is what a BDW should be* and *this is how you design and implement it* motivated our approach: an integrated, detailed, and general-purpose prescriptive contribution to design and implement BDWs, using models and methods that were adequately evaluated through different demonstration cases. Practitioners and researchers have now a set of artifacts available to be used to build BDWs and to foster future research as techniques and technologies continue to evolve.

10.1. SYNOPSIS OF THE BOOK

This book begins by presenting an overall perspective of the Big Data concept, its relevance, characteristics (pointing the several Vs and their evolution over time), and challenges, which are mainly associated with general dilemmas, the data life cycle, security and privacy concerns, and the organizational changes this concept imposes. Big

Data techniques and technologies are also highlighted, from architectural to infrastructural requirements, presenting in more detail the Lambda or NIST architectures, or describing the Hadoop ecosystem and some of its main related projects. Distributed SQL engines are also addressed, providing an overview of some existing systems, namely Drill, HAWQ, Hive, Impala, Presto, and Spark.

After this overall perspective of Big Data, this book describes in more detail several databases for Big Data environments, both for OLTP and OLAP workloads. In OLTP-oriented workloads, NoSQL databases following the key-value, column-oriented, document, or graph models are described, indicating the characteristics of each type of database and presenting technical examples for each one of them, namely using Redis, HBase, MongoDB, and Neo4j. Also, for OLTP contexts, the concept of NewSQL databases and some of their characteristics are briefly explained. In OLAP-oriented workloads, Hive is presented and described as the *de facto* SQL-on-Hadoop engine, facilitating the querying and management of large datasets that are stored in a distributed storage. For Hive, its main data organization elements – tables, partitions, and buckets – are presented, as well as the different storage formats for Hive tables (text file, sequence file, RCFile, ORC file, Avro file, and Parquet). Also, for OLAP-oriented workloads, Druid is presented as a column-oriented data store optimized for *ad hoc* interactive query processing, enhanced for grouping, filtering, and aggregating data.

Besides describing and exemplifying these databases, this book also addresses the issue of defining columnar or tabular data models in Big Data contexts, proposing specific guidelines for transforming a RDM into a CDM suitable for HBase, and for transforming a DDM into an ADM suitable for Hive (ADM_{BD}). In both cases, organizations and practitioners can take as input existing data models, with the data requirements elicited in a traditional technological environment, and move towards data models that can be implemented in a Big Data environment. In the case of an ADM, the objective is to help in the process of implementing a BDW.

In cases where the organizations and practitioners want to follow a "rip and replace" strategy for designing and implementing BDWs, this book describes an approach integrating several models and methods to build these complex systems. These models and methods were subjected to a continuous refinement process, wherein the various demonstration cases helped to facilitate the evaluation of the approach, and to iteratively improve it. The following models and methods were developed:

1. A model of logical components and data flows (section 5.2), which can be used to understand the components that should be considered in the design of a BDW, how they interoperate, and how data flows through the system. The model comprises several components related to BDW storage, processing, access, analytics, system administration and management, and security and privacy, detailing how they form a BDWing system that follows the constructs of Big Data standards such as the NBDRA.

2. A method for collecting, preparing, and enriching batch and streaming data (subsection 5.2.2), so that practitioners can understand the different steps involved from data collection to data storage in BDWing contexts. The method not only clarifies batch and streaming data flows, but also details how data science models and insights can be incorporated into data CPE workloads, enabling predictive capabilities and allowing the extraction of value from unstructured data (e.g., text, video, and image).

3. A model of the technological infrastructure (section 5.3), resulting from an extensive research and development process that took place to identify and test several technologies suitable to instantiate the different components proposed in the model of logical components and data flows. The technological infrastructure model offers various alternatives for implementing a BDWing system, including data CPE workloads, storage, querying and OLAP, data mining/machine learning, and data visualization technologies. Moreover, this model also provides some guidelines on how to deploy BDWing systems on cloud environments or on-premises.

4. A method for BDW data modeling (section 5.4), which, together with the aforementioned contributions, represents a relevant artifact to tackle one of the main problems in Big Data environments, i.e., lack of standard data modeling contributions. The method presents several constructs. These include: analytical objects; descriptive and analytical families; descriptive, factual, and predictive attributes (resulting from data science models and insights); nested attributes; granularity key; partition key; bucketing/clustering key; date, time, and spatial objects; materialized objects; complementary analytical objects; and outsourced descriptive families. The data modeling method provides a way to structure batch and streaming data using the same constructs regardless of the underlying technology supporting the storage system, providing an abstraction layer that practitioners can rely on to model BDWs supported by HDFS/Hive, NoSQL/NewSQL databases, Kudu, Druid, among other systems (see section 5.3).

The models and methods described in this book form a set of artifacts for the design and implementation of BDWing systems, and consider both structured, semi-structured, and unstructured batch and streaming data, providing adequate ways to collect, store, process, and analyze this data. These can be used by practitioners and researchers as a structured, integrated, and general-purpose approach that can be prescribed to solve several real-world BDWing problems, aiming to support structured analytics on Big Data environments while taking advantage of the BDW characteristics. Four demonstration cases helped to evaluate and enhance these models and methods.

The first demonstration case consisted in modeling six BDW data models to solve potential real-world problems. In this demonstration case, one used several artificial datasets (e.g., Adventure Works, TPC-DS, and TPC-E), as well as other publicly available datasets, including the GitHub repositories dataset, the GDELT event database, and the open-air quality API. For each dataset, a BDW data model was created using the proposed approach, which was significantly useful to thoroughly explain the several proposed models and methods. The approach in itself was developed to be relatively simple to follow, but one should take into consideration that a "rip and replace" approach like the one proposed in this book can be significantly disruptive and potentially confusing when designing and implementing specific parts of a BDWing system. Consequently, this demonstration helps to clarify when to apply specific guidelines, discussing different contexts and design decisions that practitioners may face in the future. Furthermore, this demonstration case, together with the considerations from the following data CPE demonstration case, the SSB+ Benchmark, and the SusCity demonstration case, was relevant to demonstrate the effectiveness and simplicity of our approach. This demonstration case generated several BDW data models that, avoiding complexity related to traditional dimensional DWs (e.g., different types of dimensions, bridge tables, surrogate keys, SCDs, and late arriving dimensions), take less time to structure, fuel, maintain, and extend with new batch and streaming data (structured, semi-structured, and unstructured). This inevitably results in more storage flexibility and generally more performance, and accelerates the time from data collection to analysis. Consequently, these insights offer compelling reasons for adopting a BDWing strategy in organizations.

The second demonstration case involved designing and implementing several data CPE processes focused on structured, semi-structured, and unstructured batch and streaming data, to cover different challenges related to collecting, preparing, and enriching data flowing into a BDW. Different data characteristics require different strategies, hence why this demonstration offers adequate examples to practitioners, showing the effectiveness of the proposed data CPE method. This demonstration case also highlighted the differences in complexity between traditional ETL processes and data CPE workloads, namely:

- The use of denormalized structures allows simpler processes when the underlying data source is already flat or nested, such as that generated by sensors, NoSQL databases, Excel/CSV files, XML/JSON files, Web APIs, among many others sources frequently seen in Big Data environments. Not having to develop and maintain complex workloads to fuel several types of dimension tables/concepts (e.g., mini dimensions, shrunken dimensions, junk dimensions, bridge tables, late arriving dimensions) is definitively a compelling advantage. Moreover, avoiding the need to perform constant SK lookups while loading a fact table is another advantage, especially in Big Data environments wherein one is focusing on accelerating the time to insight, instead of spending a significant amount of time trying to model and maintain the BDW. Nevertheless, if the underlying source is already relational, the contrasting phenomenon occurs, i.e., one may need to perform several join operations to fuel a certain denormalized analytical object;

- The lack of dimension tables also means that streaming scenarios are possible without complex operations like SK lookups, or complex concepts such as SCDs or late arriving dimensions. The descriptive attributes of an immutable analytical object behave like a SCD type 2 scenario, in which each record is associated with the current values of the descriptive attributes. When using outsourced descriptive families and complementary analytical objects, practitioners can consider different updating approaches, as discussed in subsections 5.4.3 and 6.2.1, to overcome some challenges similar to SCDs and late arriving dimensions.

The third demonstration case shows an extension of the SSB benchmark (O'Neil et al., 2009), the SSB+. This benchmark served to evaluate several design and implementation guidelines of the described approach, in terms of effectiveness and efficiency (e.g., latency and resource usage), using as baseline, when appropriate, a star schema DW. The results demonstrated that, generally, a fully denormalized analytical object can outperform a star schema throughout different SFs (some of them exceeding the available memory), different SQL-on-Hadoop engines, and different descriptive attributes cardinality (dimension tables size). This means that even in contexts wherein dimension tables were relatively small to fit into memory (allowing efficient map/broadcast joins), a fully denormalized analytical object was more efficient (faster execution times and less CPU usage and memory dependability), surpassing the need for constant join operations between the fact table and the corresponding dimension tables. Analytical objects were also generally faster in drill-across and window analytics scenarios. Nevertheless, this demonstration case also shows that there is space for relational structures (see subsections 5.4.3 and 5.4.4.2), which can be beneficial for reducing the storage footprint of the BDW, avoiding extreme and unnecessary redundancy and, in certain contexts, increasing processing efficiency (see subsection 8.2.3). This last insight enabled the creation of spatial, date, and time objects, and outsourced descriptive families.

Still in this demonstration case, other workloads were also evaluated in terms of query latency, including nested attributes, data partitioning, and concurrent workloads, which provided several guidelines that practitioners can take into consideration. Furthermore, the SSB+ Benchmark also served to evaluate the streaming performance of a BDW, showing that using a single query submitted through the querying and OLAP engine, one can combine batch and streaming data into a "unified picture". The streaming workload was relevant to understand the limitations of technologies like HDFS/Hive (e.g., random access disadvantages and small files problem) and NoSQL databases (e.g., Cassandra's sequential access disadvantages) when storing and retrieving vast amounts of data, and the effects they can have on query performance. These insights corroborated previous assumptions, but also complemented them with further guidelines for practitioners, which are discussed throughout Chapter 5.

Finally, the fourth demonstration case applied the approach in a smart city context, namely the SusCity research project (C. Costa & Santos, 2017c; SusCity, 2016). The SusCity BDWing system was a prototype developed in this project, in which we followed the described models and methods, proving the suitability of the BDWing approach to solve real-world problems. The architecture of the system follows the described logical components and data flows, the supporting technologies are compliant with the described technological infrastructure model, and the SusCity data model is guided by the described data modeling method. The SusCity BDW was able to support an interactive Web-based data visualization platform focusing on several smart cities concerns, such as energy, buildings efficiency, and mobility, providing adequate response times, ranging from milliseconds to a few seconds over millions of records. The SusCity data visualization platform made available several geospatial and simulation capabilities in smart cities contexts (e.g., buildings retrofitting measures and energy grid stress scenarios), in a BDW able to support new mixed and complex analytical workloads.

10.2. CONTRIBUTIONS TO THE STATE OF THE ART

According to the described work and achieved results presented above, to the best of our knowledge, this approach is a relevant contribution for the scientific and technical community, making available a set of artifacts for BDW design and implementation that not only can foster future research but, above all, can help practitioners build these complex systems, which otherwise would typically fall into a use case and *ad hoc* driven process. The models and methods described in this book were applied in four demonstration cases that allowed the evaluation of the approach mainly in terms of effectiveness, complexity, latency, and, when applicable, resource considerations (CPU usage, memory constraints, and storage footprint).

Despite the relative novelty of the topic, the approach described in this book took into consideration previously existing contributions. As such, it is built upon some general constructs and guidelines provided by the Lambda Architecture (Marz & Warren, 2015), the NBDRA (NBD-PWG, 2015), the Big Data Processing Flow (Krishnan, 2013), the Data Highway Concept (Kimball & Ross, 2013), and even some data denormalization encouragements discussed in previous works (Jukic et al., 2017; Santos et al., 2017; Santos & Costa, 2016b; Dehdouh, Bentayeb, Boussaid, & Kabachi, 2015; J. P. Costa et al., 2011). Scientific progress is often made by disruptive approaches, but it is also relevant to try to build something with a solid foundation, which was relatively difficult in this book, considering the lack of maturity and contributions related to BDWing. However, this work's contribution to the state of the art in BDWing was only possible due to previously explored paths and the relevant contributions of several related works, including the vast amounts of scientific and technical works related to traditional DWing systems, shaping several academic and professional formations, whose absence would otherwise make unfeasible the advancements regarding DWs in Big Data environments.

REFERENCES

Apache Hadoop. (2018). Welcome to Apache Hadoop. Retrieved July 3, 2018, from https://hadoop.apache.org/.

Apache Hive. (2018). Hive Transactions. Retrieved July 27, 2018, from https://cwiki.apache.org/confluence/display/Hive/Hive+Transactions.

Apache Ignite. (2018). Open source memory-centric distributed database, caching, and processing platform – Apache Ignite™. Retrieved August 7, 2018, from https://ignite.apache.org/.

Armbrust, M., Xin, R. S., Lian, C., Huai, Y., Liu, D., Bradley, J. K., ... others. (2015). Spark sql: Relational data processing in spark. In *Proceedings of the 2015 ACM SIGMOD International Conference on Management of Data* (pp. 1383–1394). ACM. Retrieved from http://dl.acm.org/citation.cfm?id=2742797.

Aslett, M. (2011). How will the database incumbents respond to NoSQL and NewSQL? The 451 Group, https://451research.com/.

Avro. (2018). Apache Avro Homepage. Retrieved December 27, 2018, from https://avro.apache.org/.

Bakshi, K. (2012). Considerations for big data: Architecture and approach. In *2012 IEEE Aerospace Conference* (pp. 1–7). https://doi.org/10.1109/AFRO.2012.6187357.

Baru, C., Bhandarkar, M., Nambiar, R., Poess, M., & Rabl, T. (2013). Benchmarking Big Data Systems and the BigData Top100 List. *Big Data*, *1*(1), 60–64. https://doi.org/10.1089/big.2013.1509.

Beauchemin, M. (2018). *Functional Data Engineering – A modern paradigm for batch data processing.* Retrieved July 23, 2019, from https://medium.com/@maximebeauchemin/functional-data-engineering-a-modern-paradigm-for-batch-data-processing-2327ec32c42a.

Begoli, E., & Horey, J. (2012). Design Principles for Effective Knowledge Discovery from Big Data. In *2012 Joint Working IEEE/IFIP Conference on Software Architecture (WICSA) and European Conference on Software Architecture (ECSA)* (pp. 215–218). https://doi.org/10.1109/WICSA-ECSA.212.32.

Boyd, D., & Crawford, K. (2012). Critical Questions for Big Data. *Information, Communication & Society*, *15*(5), 662–679. https://doi.org/10.1080/1369118X.2012.678878.

Brewer, E. (2012). CAP twelve years later: How the "rules" have changed. *Computer*, *45*(2), 23–29. https://doi.org/10.1109/MC.2012.37.

© FCA

Brown, B., Chui, M., & Manyika, J. (2011). *Are you ready for the era of 'big data'* (Report). Retrieved from http://www.t-systems.com/solutions/download-mckinsey-quarterly-/1148544_1/blobBinary/Study-McKinsey-Big-data.pdf.

Capriolo, E., Wampler, D., & Rutherglen, J. (2012). *Programming hive*. O'Reilly Media, Inc.

Cattell, R. (2011). Scalable SQL and NoSQL data stores. *ACM SIGMOD Record*, *39*(4), 12–27. https://doi.org/10.1145/1978915.1978919.

Chandarana, P., & Vijayalakshmi, M. (2014). Big Data analytics frameworks. In *2014 International Conference on Circuits, Systems, Communication and Information Technology Applications (CSCITA)* (pp. 430–434). https://doi.org/10.1109/CSCITA.2014.6839299.

Chang, F., Dean, J., Ghemawat, S., Hsieh, W. C., Wallach, D. A., Burrows, M., … Gruber, R. E. (2008). Bigtable: A Distributed Storage System for Structured Data. *ACM Trans. Comput. Syst.*, *26*(2), 4:1–4:26. https://doi.org/10.1145/1365815.1365816.

Chang, L., Wang, Z., Ma, T., Jian, L., Ma, L., Goldshuv, A., … others. (2014). HAWQ: a massively parallel processing SQL engine in hadoop. In *Proceedings of the 2014 ACM SIGMOD International Conference on Management of Data* (pp. 1223–1234). ACM. Retrieved from http://dl.acm.org/citation.cfm?id=2595636.

Chart.js. (2017). Chart.js | Open source HTML5 Charts. Retrieved March 5, 2017, from http://www.chartjs.org/.

Chen, C. L. P., & Zhang, C.-Y. (2014). Data-intensive applications, challenges, techniques and technologies: A survey on Big Data. *Information Sciences*, *275*, 314–347. https://doi.org/10.1016/j.ins.2014.01.015.

Chen, H., Chiang, R. H., & Storey, V. C. (2012). Business Intelligence and Analytics: From Big Data to Big Impact. *MIS Quarterly*, *36*(4), 1165–1188. https://doi.org/10.2307/41703503.

Chen, M., Mao, S., & Liu, Y. (2014). Big Data: A Survey. *Mobile Networks and Applications*, *19*(2), 171–209. https://doi.org/10.1007/s11036-013-0489-0.

Chodorow, K. (2013). *MongoDB: The Definitive Guide: Powerful and Scalable Data Storage* (2nd ed.). O'Reilly Media.

Chowdhury, S. (2014). *Big data and data warehouse augmentation* (Data warehouse augmentation). IBM Corporation. Retrieved from https://www.ibm.com/developerworks/analytics/library/ba-augment-data-warehouse1/ba-augment-data-warehouse1-pdf.pdf.

Clegg, D. (2015). Evolving data warehouse and BI architectures: The big data challenge. *TDWI Business Intelligence Journal*, *20*(1), 19–24.

Correia, J. (2018). *Processamento Analítico de Dados em Contextos de Big Data com o Druid*. MSc Thesis, Universidade do Minho, Portugal.

Correia, J., Santos, M. Y., Costa, C., & Andrade, C. (2018). Fast Online Analytical Processing for Big Data Warehousing. In *International Conference on Intelligent Systems (IS)* (pp. 435–442). https://doi.org/10.1109/IS.2018.8710583.

Costa, C. (2017). SSB+ GitHub Repository. Retrieved from https://github.com/epilif1017a/bigdatabenchmarks.

Costa, C., & Santos, M. Y. (2015). Improving cities sustainability through the use of data mining in a context of big city data. In *2015 International Conference of Data Mining and Knowledge*

Engineering (ICOMKE) (Vol. 1, pp. 320–325). IAENG. Retrieved from https://repositorium.sdum. uminho.pt/handle/1822/36713.

Costa, C., & Santos, M. Y. (2016a). BASIS: A big data architecture for smart cities. In *2016 SAI Computing Conference (SAI)* (pp. 1247–1256). https://doi.org/10.1109/SAI.2016.7556139.

Costa, C., & Santos, M. Y. (2016b). Reinventing the Energy Bill in Smart Cities with NoSQL Technologies. In S. Ao, G.-C. Yang, & L. Gelman (Eds.), *Transactions on Engineering Technologies* (pp. 383–396). Springer Singapore. https://doi.org/10.1007/978-981-10-1088-0_29.

Costa, C., & Santos, M. Y. (2017a). Big Data: State-of-the-art concepts, techniques, technologies, modeling approaches and research challenges. *IAENG International Journal of Computer Science*, *44*, 285–301.

Costa, C., & Santos, M. Y. (2017b). The data scientist profile and its representativeness in the European e-Competence framework and the skills framework for the information age. *International Journal of Information Management*, *37*(6), 726–734. https://doi.org/10.1016/j.ijInfomgt.2017.07.010.

Costa, C., & Santos, M. Y. (2017c). The SusCity Big Data Warehousing Approach for Smart Cities. In *Proceedings of the 21st International Database Engineering & Applications Symposium. IDEAS 2017* (pp. 264–273). ACM. https://doi.org/10.1145/3105831.3105841.

Costa, C., & Santos, M. Y. (2018). Evaluating Several Design Patterns and Trends in Big Data Warehousing Systems. In J. Krogstie & H. A. Reijers (Eds.), *Advanced Information Systems Engineering. CAISE 2018* (pp. 459–473). Springer, Cham. https://doi.org/10.1007/978-3-319-91563-0_28.

Costa, E., Costa, C., & Santos, M. Y. (2017). Efficient Big Data Modelling and Organization for Hadoop Hive-Based Data Warehouses. In M. Themistocleous & V. Morabito (Eds.), *Information Systems. EMCIS 2017* (pp. 3–16). Springer, Cham. https://doi.org/10.1007/978-3-319-65930-5_1.

Costa, E., Costa, C., & Santos, M. Y. (2018). Partitioning and Bucketing in Hive-Based Big Data Warehouses. In Á. Rocha, A. Hojjat, L. P. Reis, & S. Costanzo (Eds.), *Trends and Advances in Information Systems and Technologies. WorldCIST 2018* (pp. 764–774). Springer, Cham. https://doi.org/10.1007/978-3-319-77712-2_72.

Costa, J. P., Cecílio, J., Martins, P., & Furtado, P. (2011). ONE: A Predictable and Scalable DW Model. In *Data Warehousing and Knowledge Discovery* (pp. 1–13). Springer, Berlin, Heidelberg. https://doi.org/10.1007/978-3-642-23544-3_1.

Cuzzocrea, A., Song, I.-Y., & Davis, K. C. (2011). Analytics over large-scale multidimensional data: the big data revolution! In *Proceedings of the ACM 14th International Workshop on Data Warehousing and OLAP* (pp. 101–104). ACM. Retrieved from http://dl.acm.org/citation.cfm?id=2064695.

Dataedo. (2017). AdventureWorks – Data Dictionary. Dataedo. Retrieved from https://dataedo.com/download/AdventureWorks.pdf.

DataStax. (2018). DataStax Enterprise | Enterprise Data Management | DataStax. Retrieved December 28, 2018, from https://www.datastax.com/products/datastax-enterprise.

Davenport, T. H., Barth, P., & Bean, R. (2012). How big data is different. *MIT Sloan Management Review*, *54*(1), 43–46.

Dean, J., & Ghemawat, S. (2008). MapReduce: Simplified Data Processing on Large Clusters. *Commun. ACM*, *51*(1), 107–113. https://doi.org/10.1145/1327452.1327492.

Dehdouh, K., Bentayeb, F., Boussaid, O., & Kabachi, N. (2015). Using the column oriented NoSQL model for implementing big data warehouses. In *Proceedings of the International Conference on Parallel and Distributed Processing Techniques and Applications (PDPTA)* (p. 469). Retrieved from http://search.proquest.com/openview/fb990658e2d8b7f76720b0a6707b9e89/1?pq-origsite=gscholar.

Drill. (2018). Apache Drill – Schema-free SQL for Hadoop, NoSQL and Cloud Storage. Retrieved December 27, 2018, from https://drill.apache.org/.

Druid. (2018). Druid | Interactive Analytics at Scale. Retrieved December 27, 2018, from http://druid.io/.

Du, D. (2015). *Apache Hive Essentials*. Packt Publishing.

Dumbill, E. (2013). Making sense of big data. *Big Data*, *1*(1), 1–2. https://doi.org/10.1089/big.2012.1503.

Fan, W., & Bifet, A. (2013). Mining big data: current status, and forecast to the future. *ACM SIGKDD Explorations Newsletter*, *14*(2), 1–5. https://doi.org/10.1145/2481244.2481246.

Fisher, D., DeLine, R., Czerwinski, M., & Drucker, S. (2012). Interactions with big data analytics. *Interactions*, *19*(3), 50–59. https://doi.org/10.1145/2168931.2168943.

Floratou, A., Minhas, U. F., & Özcan, F. (2014). SQL-on-Hadoop: Full Circle Back to Shared-nothing Database Architectures. *Proc. VLDB Endow.*, *7*(12), 1295–1306. https://doi.org/10.14778/2732977.2733002.

Gai, K., Qiu, M., & Sun, X. (2018). A survey on FinTech. *Journal of Network and Computer Applications*, *103*, 262–273. https://doi.org/10.1016/j.jnca.2017.10.011.

Gandomi, A., & Haider, M. (2015). Beyond the hype: Big data concepts, methods, and analytics. *International Journal of Information Management*, *35*(2), 137–144. https://doi.org/10.1016/j.ijinfomgt.2014.10.007.

Garber, L. (2012). Using In-Memory Analytics to Quickly Crunch Big Data. *Computer*, *45*(10), 16–18. https://doi.org/10.1109/MC.2012.358.

Gates, A. (2014). Stinger.next: Enterprise SQL at Hadoop Scale with Apache Hive. Retrieved January 25, 2019, from https://hortonworks.com/blog/stinger-next-enterprise-sql-hadoop-scale-apache-hive/.

GDELT. (2018). The GDELT Project. Retrieved August 6, 2018, from https://www.gdeltproject.org/.

George, L. (2010). *HBase: The Definitive Guide* (2nd ed.). O'Reilly Media, Inc.

Ghemawat, S., Gobioff, H., & Leung, S.-T. (2003). The Google file system. In *ACM SIGOPS operating systems review* (Vol. 37, pp. 29–43). ACM. https://doi.org/10.1145/1165389.945450.

Google. (2018). GitHub Repositories Dataset on Google BigQuery. Retrieved August 6, 2018, from https://bigquery.cloud.google.com/dataset/bigquery-public-data:github_repos.

Google Maps. (2017). Google Maps JavaScript API. Retrieved March 5, 2017, from https://developers.google.com/maps/documentation/javascript/.

Google Trends. (2018). Interest in Big Data over time. Retrieved August 7, 2018, from https://www.google.pt/trends/explore#q=big%20data.

Grolinger, K., Higashino, W. A., Tiwari, A., & Capretz, M. A. (2013). Data management in cloud environments: NoSQL and NewSQL data stores. *Journal of Cloud Computing: Advances, Systems and Applications*, *2*(1), 22.

Grover, A., Gholap, J., Janeja, V. P., Yesha, Y., Chintalapati, R., Marwaha, H., & Modi, K. (2015). SQL-like big data environments: Case study in clinical trial analytics. In *2015 IEEE International Conference on Big Data (Big Data)* (pp. 2680–2689). https://doi.org/10.1109/BigData.2015.7364068.

Hall, M., Frank, E., Holmes, G., Pfahringer, B., Reutemann, P., & Witten, I. H. (2009). The WEKA data mining software: an update. *ACM SIGKDD Explorations Newsletter, 11*(1), 10–18.

Han, J., Pei, J., & Kamber, M. (2012). *Data Mining: Concepts and Techniques* (3rd ed.). Elsevier.

Hashem, I. A. T., Yaqoob, I., Anuar, N. B., Mokhtar, S., Gani, A., & Khan, S. U. (2015). The rise of "big data" on cloud computing: Review and open research issues. *Information Systems, 47*, 98–115. https://doi.org/10.1016/j.is.2014.07.006.

Hausenblas, M., & Nadeau, J. (2013). Apache drill: interactive ad-hoc analysis at scale. *Big Data, 1*(2), 100–104.

HAWQ. (2018). Apache HAWQ® Homepage. Retrieved December 27, 2018, from http://hawq.apache.org/.

HBase. (2018a). Apache HBase – Apache HBase™ Home. Retrieved June 22, 2018, from https://hbase.apache.org/.

HBase. (2018b). Apache HBase™ Reference Guide. Retrieved December 27, 2018, from https://hbase.apache.org/2.0/book.html.

Hevner, A. R., March, S. T., Park, J., & Ram, S. (2004). Design Science in Information Systems Research. *MIS Q., 28*(1), 75–105.

Hewitt, E. (2010). *Cassandra: The Definitive Guide* (1st ed.). Beijing: O'Reilly Media.

Hive. (2018a). Apache Hive™. Retrieved December 27, 2018, from https://hive.apache.org/.

Hive. (2018b). LanguageManual ORC – Apache Hive – Apache Software Foundation. Retrieved December 27, 2018, from https://cwiki.apache.org/confluence/display/Hive/LanguageManual+ORC.

Hive. (2018c). RCFile – Apache Hive – Apache Software Foundation. Retrieved December 27, 2018, from https://cwiki.apache.org/confluence/display/Hive/RCFile.

Hortonworks. (2016). *Solving Apache Hadoop Security: A Holistic Approach to a Secure Data Lake* (White paper). Hortonworks. Retrieved from http://hortonworks.com/info/solving-hadoop-security/.

Huai, Y., Chauhan, A., Gates, A., Hagleitner, G., Hanson, E. N., O'Malley, O., … Zhang, X. (2014). Major Technical Advancements in Apache Hive. In *Proceedings of the 2014 ACM SIGMOD International Conference on Management of Data* (pp. 1235–1246). ACM. https://doi.org/10.1145/2588555.2595630.

IBM. (2018). IBM Db2 – Data management software – IBM Analytics. Retrieved December 28, 2018, from https://www.ibm.com/analytics/us/en/db2/.

Impala. (2018). Apache Impala Homepage. Retrieved December 27, 2018, from https://impala.apache.org/.

Inmon, W. H., & Linstedt, D. (2014). *Data Architecture: A Primer for the Data Scientist: Big Data, Data Warehouse and Data Vault* (1st ed.). Morgan Kaufmann.

Intel IT Center. (2012). *Peer Research: Big Data Analytics* (Report). Intel. Retrieved from http://www.intel.com/content/dam/www/public/us/en/documents/reports/data-insights-peer-research-report.pdf.

Jagadish, H. V., Gehrke, J., Labrinidis, A., Papakonstantinou, Y., Patel, J. M., Ramakrishnan, R., & Shahabi, C. (2014). Big data and its technical challenges. *Communications of the ACM*, *57*(7), 86–94. http://dx.doi.org/10.1145/2611567.

Jara, A. J., Bocchi, Y., & Genoud, D. (2013). Determining human dynamics through the internet of things. In *Proceedings of the 2013 IEEE/WIC/ACM International Joint Conferences on Web Intelligence (WI) and Intelligent Agent Technologies (IAT)-Volume 03* (pp. 109–113). IEEE Computer Society. Retrieved from http://dl.acm.org/citation.cfm?id=2569254.

Jethro. (2018). Hadoop Hive and 11 SQL-on-Hadoop Alternatives. Retrieved December 27, 2018, from https://jethro.io/hadoop-hive.

Ji, C., Li, Y., Qiu, W., Awada, U., & Li, K. (2012). Big Data Processing in Cloud Computing Environments. *Proceedings of the 2012 12th International Symposium on Pervasive Systems, Algorithms, and Networks (i-Span 2012)*, 17–23. https://doi.org/10.1109/I-SPAN.2012.9.

jQuery. (2017). jQuery Homepage. Retrieved March 5, 2017, from https://jquery.com/.

Jukic, N., Jukic, B., Sharma, A., Nestorov, S., & Korallus Arnold, B. (2017). Expediting analytical databases with columnar approach. *Decision Support Systems*, *95*, 61–81. https://doi.org/10.1016/j.dss.2016.12.002.

Kafka. (2018). Apache Kafka Homepage. Retrieved July 3, 2018, from https://kafka.apache.org/.

Kambatla, K., Kollias, G., Kumar, V., & Grama, A. (2014). Trends in big data analytics. *Journal of Parallel and Distributed Computing*, *74*(7), 2561–2573. https://doi.org/10.1016/j.jpdc.2014.01.003.

Khurana, A. (2012). Introduction to HBase schema design. *Login*, *37*(5), 29–36.

Kimball, R., & Ross, M. (2013). *The data warehouse toolkit: The definitive guide to dimensional modeling* (3rd ed.). John Wiley & Sons.

Kornacker, M., Behm, A., Bittorf, V., Bobrovytsky, T., Choi, A., Erickson, J., … Yoder, M. (2015). Impala: A modern, open-source sql engine for hadoop. In *7th Biennial Conference on Innovative Data Systems Research (CIDR'15)*.

Krishnan, K. (2013). *Data Warehousing in the Age of Big Data* (1st ed.). San Francisco, CA, USA: Morgan Kaufmann Publishers Inc.

Landset, S., Khoshgoftaar, T. M., Richter, A. N., & Hasanin, T. (2015). A survey of open source tools for machine learning with big data in the Hadoop ecosystem. *Journal of Big Data*, *2*(1), 24. https://doi.org/10.1186/s40537-015-0032-1.

Laney, D. (2001). *3D Data Management: Controlling Data Volume, Velocity, and Variety* (Report). META Group Inc. Retrieved from http://blogs.gartner.com/doug-laney/files/2012/01/ad949-3D-Data-Management-Controlling-Data-Volume-Velocity-and-Variety.pdf.

LaValle, S., Lesser, E., Shockley, R., Hopkins, M. S., & Kruschwitz, N. (2011). Big data, analytics and the path from insights to value. *MIT Sloan Management Review*, *52*(2), 21–32.

Lipcon, T., Alves, D., Burkert, D., Cryans, J., Dembo, A., Percy, M., … McCabe, C. P. (2015). *Kudu: Storage for Fast Analytics on Fast Data*. Cloudera. Retrieved from http://getkudu.io/kudu.pdf.

Mackey, G., Sehrish, S., & Wang, J. (2009). Improving metadata management for small files in HDFS. In *2009 IEEE International Conference on Cluster Computing and Workshops* (pp. 1–4). https://doi.org/10.1109/CLUSTR.2009.5289133.

Madden, S. (2012). From databases to big data. *IEEE Internet Computing*, *16*(3), 4–6. http://dx.doi.org/10.1109/MIC.2012.50.

Manyika, J., Chui, M., Brown, B., Bughin, J., Dobbs, R., Roxburgh, C., & Byers, A. H. (2011). *Big data: The next frontier for innovation, competition, and productivity* (Report). McKinsey Global Institute. Retrieved from http://www.citeulike.org/group/18242/article/9341321.

MapR. (2018). SQL on Hadoop Details. Retrieved December 27, 2018, from https://mapr.com/why-hadoop/sql-hadoop/sql-hadoop-details/.

Marz, N., & Warren, J. (2012). *Big Data: Principles and best practices of scalable realtime data systems* (1st ed.). Manning Publications Co.

Marz, N., & Warren, J. (2015). *Big Data: Principles and best practices of scalable realtime data systems* (2nd ed.). Manning Publications Co.

McAfee, A., Brynjolfsson, E., Davenport, T. H., Patil, D. J., & Barton, D. (2012). Big data. The Management Revolution. *The Management Revolution. Harvard Bus Rev, 90*(10), 61–67.

Melnik, S., Gubarev, A., Long, J. J., Romer, G., Shivakumar, S., Tolton, M., & Vassilakis, T. (2010). Dremel: interactive analysis of web-scale datasets. *Proceedings of the VLDB Endowment, 3*(1–2), 330–339.

MemSQL. (2018). MemSQL is the No-Limits Database Powering Modern Applications and Analytical Systems. Retrieved December 28, 2018, from https://www.memsql.com/.

Michael, K., & Miller, K. W. (2013). Big Data: New Opportunities and New Challenges [Guest editors' introduction]. *Computer, 46*(6), 22–24. https://doi.org/10.1109/MC.2013.196.

Microsoft. (2018a). SQL Server 2016 | Microsoft. Retrieved December 28, 2018, from https://www.microsoft.com/en-us/sql-server/sql-server-2016.

Microsoft. (2018b). *SQL-server-samples: Official Microsoft GitHub Repository containing code samples for SQL Server*. Microsoft. Retrieved from https://github.com/Microsoft/sql-server-samples.

MongoDB. (2018). Open Source Document Database. Retrieved December 27, 2018, from https://www.mongodb.com/index.

Monteiro, C. S., Costa, C., Pina, A., Santos, M. Y., & Ferrão, P. (2018). An urban building database (UBD) supporting a smart city information system. *Energy and Buildings, 158*, 244–260. https://doi.org/10.1016/j.enbuild.2017.10.009.

NBD-PWG. (2015). *NIST Big Data Interoperability Framework: Volume 6, Reference Architecture* (Technical Report No. NIST SP 1500-6). National Institute of Standards and Technology. Retrieved from http://nvlpubs.nist.gov/nistpubs/SpecialPublications/NIST.SP.1500-6.pdf.

Neo4j. (2018). Neo4j Graph Platform – The Leader in Graph Databases. Retrieved December 27, 2018, from https://neo4j.com/.

NoSQL. (2018). NOSQL Databases. Retrieved November 1, 2018, from http://nosql-database.org/.

O'Neil, P. E., O'Neil, E. J., & Chen, X. (2009). *The star schema benchmark (SSB)*. Retrieved from http://www.cs.umb.edu/~poneil/StarSchemaB.PDF

OpenAQ. (2018). OpenAQ Platform. Retrieved August 6, 2018, from https://openaq.org/.

Oracle. (2018). In-Memory Database | Oracle. Retrieved December 28, 2018, from https://www.oracle.com/database/technologies/in-memory.html.

Parquet. (2018). Apache Parquet Homepage. Retrieved March 27, 2016, from https://parquet.apache.org/.

Pavlo, A., & Aslett, M. (2016). What's Really New with NewSQL? *SIGMOD Rec.*, *45*(2), 45–55. https://doi.org/10.1145/3003665.3003674.

Peffers, K., Tuunanen, T., Rothenberger, M., & Chatterjee, S. (2007). A Design Science Research Methodology for Information Systems Research. *J. Manage. Inf. Syst.*, *24*(3), 45–77. https://doi.org/10.2753/MIS0742-1222240302.

Phoenix. (2018). Overview | Apache Phoenix. Retrieved December 27, 2018, from https://phoenix.apache.org/.

Pivotal. (2017, January 17). Apache MADlib [text/html]. Retrieved December 27, 2018, from https://pivotal.io/madlib.

Presto. (2018). Presto Homepage. Retrieved March 18, 2018, from https://prestodb.io.

Provost, F., & Fawcett, T. (2013). Data Science and its Relationship to Big Data and Data-Driven Decision Making. *Big Data*, *1*(1), 51–59. https://doi.org/10.1089/big.2013.1508.

Pujari, A. K. (2001). *Data mining techniques* (1st ed.). Universities press.

Qiao, L., Li, Y., Takiar, S., Liu, Z., Veeramreddy, N., Tu, M., … others. (2015). Gobblin: Unifying data ingestion for Hadoop. *Proceedings of the VLDB Endowment*, *8*(12), 1764–1769.

Redis. (2018a). EBOOK – Redis in Action. Retrieved December 27, 2018, from https://redislabs.com/ebook/part-1-getting-started/chapter-1-getting-to-know-redis/1-2-what-redis-data-structures-look-like/.

Redis. (2018b). Redis Homepage. Retrieved August 7, 2018, from https://redis.io/.

Robinson, I., Webber, J., & Eifrem, E. (2015). *Graph Databases: New Opportunities for Connected Data* (2 edition). O'Reilly Media.

Rodrigues, M., Santos, M. Y., & Bernardino, J. (2018). Big data processing tools: An experimental performance evaluation. *Wiley Interdisciplinary Reviews: Data Mining and Knowledge Discovery*, e1297. https://doi.org/10.1002/widm.1297.

Russom, P. (2011). *Big data analytics* (Best Practices Report) (pp. 1–35). TDWI Research. Retrieved from https://tdwi.org/~/media/0C630BCFD9064A9287148F1FA33460E4.pdf.

Russom, P. (2014). *Evolving Data Warehouse Architectures in the Age of Big Data* (TDWI Best Practices Report). The Data Warehouse Institute. Retrieved from https://tdwi.org/research/2014/04/best-practices-report-evolving-data-warehouse-architectures-in-the-age-of-big-data.aspx.

Russom, P. (2016). *Data Warehouse Modernization in the Age of Big Data Analytics* (TDWI Best Practices Report). The Data Warehouse Institute. Retrieved from https://tdwi.org/research/2016/03/best-practices-report-data-warehouse-modernization/asset.aspx?tc=assetpg.

Sadalage, P. J., & Fowler, M. (2012). *NoSQL Distilled: A Brief Guide to the Emerging World of Polyglot Persistence* (1st ed.). Addison-Wesley Professional.

Sagiroglu, S., & Sinanc, D. (2013). Big data: A review. In *2013 International Conference on Collaboration Technologies and Systems (CTS)* (pp. 42–47). https://doi.org/10.1109/CTS.2013.6567202.

Santos, M. Y., & Costa, C. (2016a). Data Models in NoSQL Databases for Big Data Contexts. In *2016 International Conference on Data Mining and Big Data (DMBD)* (pp. 1–11). Springer-Verlag, LNCS 9714. https://doi.org/10.1007/978-3-319-40973-3_48.

Santos, M. Y., & Costa, C. (2016b). Data Warehousing in Big Data: From Multidimensional to Tabular Data Models. In *Proceedings of the Ninth International C* Conference on Computer Science & Software Engineering* (pp. 51–60). ACM. https://doi.org/10.1145/2948992.2949024.

Santos, M. Y., Costa, C., Galvão, J., Andrade, C., Martinho, B. A., Lima, F. V., & Costa, E. (2017). Evaluating SQL-on-Hadoop for Big Data Warehousing on Not-So-Good Hardware. In *Proceedings of the 21st International Database Engineering & Applications Symposium. IDEAS 2017* (pp. 242–252). ACM. https://doi.org/10.1145/3105831.3105842.

Santos, M. Y., & Ramos, I. (2009). *Business Intelligence: Tecnologias da informação na gestão de conhecimento* (2nd ed.). Lisboa: FCA – Editora de Informática.

SAP. (2018). What is SAP HANA | In Memory Computing and Real Time Analytics. Retrieved December 28, 2018, from https://www.sap.com/products/hana.html.

Schroeck, M., Shockley, R., Janet, S., Romero-Morales, D., & Tufano, P. (2012). Analytics: The real-world use of big data. How innovative enterprises extract value from uncertain data. IBM Institute for Business Value. Retrieved from https://www.ibm.com/smarterplanet/global/files/se__sv_se__intelligence__Analytics_-_The_real-world_use_of_big_data.pdf.

Shanahan, J. G., & Dai, L. (2015). Large scale distributed data science using apache spark. In *Proceedings of the 21th ACM SIGKDD International Conference on Knowledge Discovery and Data Mining* (pp. 2323–2324). ACM. Retrieved from http://dl.acm.org/citation.cfm?id=2789993.

Shanklin, C. (2017, May 11). Ultra-fast OLAP Analytics with Apache Hive and Druid. Retrieved December 27, 2018, from https://hortonworks.com/blog/apache-hive-druid-part-1-3/.

Shvachko, K., Kuang, H., Radia, S., & Chansler, R. (2010). The Hadoop Distributed File System. In *2010 IEEE 26th Symposium on Mass Storage Systems and Technologies (MSST)* (pp. 1–10). https://doi.org/10.1109/MSST.2010.5496972.

Soliman, M. A. (2017). Big Data Query Engines. In A. Y. Zomaya & S. Sakr (Eds.), *Handbook of Big Data Technologies* (pp. 179–217). Springer, Cham. https://doi.org/10.1007/978-3-319-49340-4_6.

Spark. (2017). Spark SQL and DataFrames – Spark 2.1.1 Documentation. Retrieved May 30, 2017, from https://spark.apacho.org/docs/latest/sql-programming-guide.html#datasets-and-dataframes.

Spark. (2018). Apache Spark™ – Unified Analytics Engine for Big Data. Retrieved March 18, 2018, from https://spark.apache.org/.

SusCity. (2016). SUSCITY – An MIT Portugal Project. Retrieved May 4, 2016, from http://suscity-project.eu/inicio/.

Szegedi, I. (2014). Pivotal Hadoop Distribution and HAWQ Realtime Query Engine. Retrieved January 25, 2019, from https://dzone.com/articles/pivotal-hadoop-distribution.

Talend. (2017). Talend Open Studio for Big Data Product Details. Retrieved March 5, 2017, from https://www.talend.com/download_page_type/talend-open-studio/.

Thusoo, A., Sarma, J. S., Jain, N., Shao, Z., Chakka, P., Zhang, N., … Murthy, R. (2010). Hive-a petabyte scale data warehouse using hadoop. In *IEEE 26th International Conference on Data Engineering (ICDE)* (pp. 996–1005). IEEE. Retrieved from http://ieeexplore.ieee.org/xpls/abs_all.jsp?arnumber=5447738.

© FCA

Tien, J. M. (2013). Big Data: Unleashing information. *Journal of Systems Science and Systems Engineering*, *22*(2), 127–151. https://doi.org/10.1007/s11518-013-5219-4.

TPC. (2017a). TPC-DS Homepage. Retrieved August 16, 2017, from http://www.tpc.org/tpcds/.

TPC. (2017b). TPC-H Homepage. Retrieved August 16, 2017, from http://www.tpc.org/tpch/.

TPC. (2018). TPC-E Homepage. Retrieved August 3, 2018, from http://www.tpc.org/tpce/.

Tudorica, B. G., & Bucur, C. (2011). A comparison between several NoSQL databases with comments and notes. In *Roedunet International Conference (RoEduNet), 2011 10th* (pp. 1–5). IEEE. Retrieved from http://ieeexplore.ieee.org/xpls/abs_all.jsp?arnumber=5993686.

Vale Lima, F., Costa, C., & Santos, M. Y. (2019). Real-Time Big Data Warehousing. In D. Taniar & W. Rahayu (Eds.), *Emmerging Perspectives in Big Data Warehousing* (pp. 28–57). https://doi.org/10.4018/978-1-5225-5516-2.

Vieira, A. A. C., Pedro, L., Santos, M. Y., Fernandes, J. M., & Dias, L. S. (2019). Data Requirements Elicitation in Big Data Warehousing. In Marinos Themistocleous & P. Rupino da Cunha (Eds.), *Information Systems* (pp. 106–113). Springer International Publishing.

Vijayakumar, V., & Nedunchezhian, R. (2012). A study on video data mining. *International Journal of Multimedia Information Retrieval*, *1*(3), 153–172. https://doi.org/10.1007/s13735-012-0016-2.

Villars, R. L., Olofson, C. W., & Eastwood, M. (2011). *Big data: What it is and why you should care* (Report). IDC. Retrieved from http://www.tracemyflows.com/uploads/big_data/idc_amd_big_data_whitepaper.pdf

VoltDB. (2018). In-Memory Database. Retrieved December 27, 2018, from https://www.voltdb.com/.

Wang, H., Qin, X., Zhang, Y., Wang, S., & Wang, Z. (2011). LinearDB: A Relational Approach to Make Data Warehouse Scale Like MapReduce. In J. X. Yu, M. H. Kim, & R. Unland (Eds.), *Database Systems for Advanced Applications* (pp. 306–320). Springer, Berlin, Heidelberg. https://doi.org/10.1007/978-3-642-20152-3_23.

Ward, J. S., & Barker, A. (2013). Undefined By Data: A Survey of Big Data Definitions. *ArXiv:1309.5821 [Cs.DB]*. Retrieved from http://arxiv.org/abs/1309.5821.

White, T. (2015). *Hadoop: The Definitive Guide* (4th ed.). O'Reilly Media.

Wigan, M. R., & Clarke, R. (2013). Big Data's Big Unintended Consequences. *Computer*, *46*(6), 46–53. https://doi.org/10.1109/MC.2013.195.

Xu, W., Luo, W., & Woodward, N. (2012). Analysis and Optimization of Data Import with Hadoop. In *2012 IEEE 26th International Parallel and Distributed Processing Symposium Workshops PhD Forum* (pp. 1058–1066). https://doi.org/10.1109/IPDPSW.2012.129.

Yang, F., Tschetter, E., Léauté, X., Ray, N., Merlino, G., & Ganguli, D. (2014). Druid: a real-time analytical data store (pp. 157–168). ACM Press. https://doi.org/10.1145/2588555.2595631.

Yuhanna, N., & Gualtieri, M. (2016). *Emerging Technology: Translytical Databases Deliver Analytics At The Speed Of Transactions*. Forrester Research. Retrieved from http://www.odbms.org/wp-content/uploads/2017/10/Forrester-report-Translytical-Databases.pdf.

Zhang, J., Hsu, W., & Lee, M. L. (2001). Image Mining: Issues, Frameworks and Techniques. In *Proceedings of the Second International Conference on Multimedia Data Mining* (pp. 13–20). Springer-Verlag. Retrieved from http://dl.acm.org/citation.cfm?id=3012377.3012378.

Zikopoulos, P., & Eaton, C. (2011). *Understanding Big Data: Analytics for Enterprise Class Hadoop and Streaming Data* (1st ed.). McGraw-Hill Osborne Media.

INDEX

© FCA